Cambridge Computer Science Texts · 9

An Introduction to Computational Combinatorics

E. S. PAGE

Vice-Chancellor
University of Reading

L. B. WILSON

Computing Laboratory
University of Newcastle upon Tyne

Cambridge University Press

Cambridge
London · New York · Melbourne

Published by the Syndics of the Cambridge University Press

The Pitt Building, Trumpington Street, Cambridge CB2 1RP

Bentley House, 200 Euston Road, London NW1 2DB

32 East 57th Street, New York, NY 10022, USA

296 Beaconsfield Parade, Middle Park, Melbourne 3206, Australia

© Cambridge University Press 1979

First published 1979

Reproduced, printed and bound in Great Britain by

Cox & Wyman Ltd, London, Fakenham and Reading

ISBN 0 521 22427 6 hard covers

ISBN 0 521 29492 4 paperback

Contents

		Page
Preface		v
Chapter 1 The Problems of Computational Combinatorics		1
1.1	Introduction	1
1.2	Where?	1
1.3	Which is Best?	3
1.4	Methods	4
Chapter 2 Constant Coefficient Difference Equations		5
2.1	Introduction	5
2.2	The Homogeneous Equation	8
2.3	The Complete Equation	11
2.4	Solution by Generating Functions	23
2.5	Simultaneous Equations	25
2.6	Applications	26
2.7	Bibliography	30
2.8	Exercises	31
Chapter 3 Other Difference Equations		36
3.1	Variable Coefficients	36
3.2	Reduction of the Order	42
3.3	Partial Difference Equations	48
3.4	Bibliography	51
3.5	Exercises	52
Chapter 4 Elementary Configurations		54
4.1	Introduction	54
4.2	Combinations and Permutations	54
4.3	Generating Functions	60
4.4	The Principle of Inclusion and Exclusion	64
4.5	Partitions and Compositions of Integers	67
4.6	Graphs and Trees	74

4.7	Bibliography	88
4.8	Exercises	90
Chapter 5	**Ordering and Generation of Elementary Configurations**	97
5.1	Lexicographical Orderings	97
5.2	Permutations	103
5.3	The Generation of Combinations	116
5.4	The Generation of Compositions	118
5.5	The Generation of Partitions	120
5.6	Recurrence Relations	122
5.7	Bibliography	125
5.8	Exercises	127
Chapter 6	**Search Procedures**	131
6.1	Backtrack Programming	131
6.2	Branch and Bound Methods	139
6.3	Dynamic Programming	149
6.4	Complexity	154
6.5	Bibliography	156
6.6	Exercises	157
Chapter 7	**Theorems and Algorithms for Selection**	164
7.1	Introduction	164
7.2	Systems of Distinct Representatives	165
7.3	Algorithms for Systems of Distinct Representatives	169
7.4	Matrices of Zeros and Ones	173
7.5	Assignment Problems	175
7.6	Bibliography	188
7.7	Exercises	189
	Notes on the Solutions to Exercises	195
	Index	215

Preface

By the time students have done some programming in one or two languages and have learnt the common ways of representing information in a computer they may wish to embark upon further study of theoretical or applied topics in computing science. Most of them will encounter problems which need one or more of the techniques described in this book; for example, the analyses of certain algorithms and of some models of scheduling strategies in an operating system depend upon the formation and solution of difference equations; the tasks of making lists of possible alternatives and of answering questions about them crop up in as diverse areas as stock records and the theory of grammars; searches for discrete optima - or for the best that can be found with just so much computing - occur in manufacturing, design and many operational research investigations. These examples would be justification enough for the teaching of this material but we believe that the field of computational combinatorics itself contains fascinating problems and we hope that this introduction gives a glimpse of some of them.

It will be obvious from a count of the numbered equations that some chapters are more mathematical than others, and that chapters 2 to 4 have a higher 'equation count' than the rest. We could almost as easily have changed the order of presentation and put this material near the end, and some readers and lecturers will prefer to do so. However, mathematical techniques have to be acquired if one is to calculate how large or lengthy a combinatorial calculation will be. Problems can grow so big so quickly that much human and computer time, and money, can be wasted if an infeasible computation is started. We therefore put this material first in our courses but it could be postponed and the more practical programming topics treated first instead, referring to chapters 2 to 4 for definitions where necessary.

It is convenient for us to teach most of the material in this book during the second year of certain groups of students at Newcastle; but the mathematical and computing science preparation that students have received will vary from place to place and elsewhere other times might be more appropriate. The mathematics needed is elementary algebra and calculus - for example, an isolated use of Taylor's theorem occurs in chapter 2 and simple first order differential equations in chapter 3 - and the students should have sufficient computing experience to be able to understand the algorithms which are described here in an Algol-like language, and to program them for running on the machine available to them. The students can check their grasp of what they have read by tackling the exercises given at the end of the chapters; some of them are straightforward drill but others are longer, taken from university examination papers, which will give an idea of the standard required in a written paper at this level. There are also suggestions for some programming tasks which can be expanded if the time and course structure so demand. Some hints on the solutions of the exercises are given at the end of the book; we hope that they are sufficiently detailed to indicate the way to proceed to someone who is stuck but not so full as to be specimen answers which a student might find attractive to copy.

We are pleased to acknowledge our gratitude to the Universities of Newcastle upon Tyne, Warwick and Linkoping for their permission to include questions from their examination papers.

However hard authors try to eliminate all errors and misprints they will be fortunate if they succeed in doing so; if any remain in spite of our own efforts and the helpful comments of some of our colleagues we apologise.

E. S. Page
L. B. Wilson

1 · The Problems of Computational Combinatorics

1.1 INTRODUCTION

Digital computers represent the items which they store and manipulate in a discrete form. The operations that are performed on the items are exact and they are necessarily finite in number even though each operation is completed very quickly. Each item may need only a few bits or bytes for its storage but there may be many such items. When one plans a computer application one normally needs to know, at least approximately, how much storage will be required and about how big a computation it is. A major component of the storage needed may be the number of items of a particular type that have to be stored. Similarly, a knowledge of how many operations of the various types the computation involves will help in assessing how much computing will be needed. These quantities, both dependent upon the answer to a question 'How many?', are important if one attempts to compare different methods of achieving the desired end or even to decide whether the computation is feasible. In computing some use of one resource of the computing system can nearly always be reduced at the expense of making use of another of the resources; for example, there can nearly always be a trade of storage for processing time, and vice versa. For a proper comparison of the alternative methods the amounts need to be quantified. One needs to be able to answer the questions which start 'How many?' - one of the interests included in the subject of computational combinatorics.

1.2 WHERE?

For some applications in computing, the answers to questions of the 'How many?' type may be sufficient; for example, if two algorithms which each perform a certain computation are being compared it may be enough to know how many operations of the various types each algorithm requires and how many storage locations each needs. In some other applications an answer to the question 'How many?', whether approximate or exact, may only be enough to determine whether or not the problem can be tackled at all. For example, the task of sorting some items into order within the main storage of the machine will only be possible by the given algorithm if the number of items is small enough so that the storage needed for the final list and for intermediate working storage is within the bounds of what can be spared for this program. However, once an ordered list has been obtained, other questions may need answering. Does a certain item appear in that ordered list, and, if so, where is it? Such a question might call for a suitable searching algorithm but it might be possible to answer it by calculation without searching the list itself. The converse question 'Which item is at a given position in that order?' could present a task of retrieval but it might be possible to identify and construct the item without referring to the order by performing an appropriate computation. The latter approach would be the only feasible one if the number of items were so large that not all could be conveniently stored and also if the items had not been explicitly generated and it was wasteful to do so. For example, the number of permutations of n items increases rapidly with n and soon becomes too large to generate and store (there are over 400 million even for n = 12) but there are several ways of defining an order so that the questions can be answered directly. In other problems answers to 'Where?' and 'Which?' questions may be found most suitably by applying one of the number of techniques available for searching and retrieval, but there can be cases where such techniques are not applicable or can be replaced by more efficient ones special to the application. In a sense, all these problems are included in computational combinatorics but as algorithms for searching, sorting and retrieval are normally introduced at an early stage of study in computing science the latter are only mentioned incidentally in this book.

The problem of sorting - given a set of items, put them in order - has a counterpart when the items are not given explicitly, but instead are defined in some implicit manner. For example, the set of items may be all possible arrangements of the four letters A, B, C, D. In such a case, the problem is to generate the set as well as to place the items in order, but it is often possible to choose the method of generation which will produce the desired order directly, and also enable other questions about the order to be answered easily.

1.3 **WHICH IS BEST?**

Problems of optimisation have claimed the attention of mathematicians from earliest times; some appear elementary - like finding the shortest distance from a point to a given circle - others, while calling for more advanced methods, are routine - like the 'soup tin' problem of obtaining the cylinder of greatest volume for a given surface area, or discovering the closed curve of given length which encloses the greatest area. Yet others are comparatively simple in specific cases but difficult in general - the task of showing that at most four paints are needed to colour a given planar map is of a different type and order of difficulty from proving that all such maps can be coloured by at most four paints. The problems of optimisation encountered in computational combinatorics are, not surprisingly, concerned with discrete rather than continuous variables, and are predominantly of the apparently trivial type 'Which is the best item out of this finite set of items?'. Of course, the method by which one item is compared with another may be complicated but the principal difficulty with such problems is that the set concerned, though finite, is large. Indeed, it may be so large that an exhaustive search through it will be beyond the resources of the computing equipment available. Thus we look for algorithms better than mere exhaustive searching. For example, we know that there are only (!) n! different ways of visiting n towns once and once only starting from another given town; one of these routes must have the shortest length and so it must be easy to discover it and so solve this 'travelling salesman' problem: so it must, were it not for the vastness of n! for n more than a few handfuls. A naive approach soon founders. We need algorithms which use the properties of the set and of the criteria determining the optimum to reduce the size of the computation. The construction of such algorithms and the assessment of the work they entail is another of the topics studied in computational combinatorics.

A more theoretical aspect, but one often with great practical significance, is that which describes attempts to quantify how 'difficult' classes of computations are. This field of study, called computational complexity, assesses how the numbers of operations of various types in the best possible algorithms increase with the size of the problem. Sometimes the study yields a constructive proof leading to a practical algorithm; in other cases, the most that can be done is to show the equivalence of different classes of (rather difficult) problems and to conjecture how fast the number of operations required increases with the size of the problem. We restrict ourselves to the mention of one or two problems of this kind.

1.4 **METHODS**

Readers will recall from their studies in elementary algebra the method of proof by mathematical induction. In a simple form this requires one to demonstrate that if a proposition is true for a general value of some integer parameter n, it is also true for the next larger integer, $n + 1$, and then to exhibit a case for a particular value of n, usually a small one ($n = 1$ or 2 perhaps), in which the proposition is true. It then follows that the proposition is true for any greater integer value of n quite generally. The idea of relating the situation for one value of n with other, usually adjacent, values (say $n - 1$ and $n - 2$) is one which is used widely in combinatorial problems. These recurrence relations when they can be found are most useful tools in many problems in computational combinatorics. Sometimes they can be solved as difference equations to answer questions of the 'How many?' kind; in other examples they can be used as the basis of algorithms to compute 'How many?' for specific cases or to generate lists and solutions for bigger problems from smaller ones that have already been solved, and to exhibit the structure of problems and to reveal relations between different problems. The methods which are introduced in this book first show how to solve some of the important types of difference equations and how to apply them to problems in computational combinatorics. Recurrence relations are used to develop various algorithms, and appear frequently in different guises - sometimes a generating function is used to solve the corresponding difference equation, while at other times a generating function gives a means of deriving a recurrence relation. The recurrence relation itself provides some unity over a field which includes a great variety of different kinds of problems.

2 · Constant Coefficient Difference Equations

2.1 INTRODUCTION

Many problems which seek to determine how many objects of a particular type there are, can be made to depend upon a single variable which takes integer values; these values are often the natural numbers $0, 1, 2, 3, 4, \ldots$ but sometimes they may be just a subset like the even positive integers, $2, 4, 6, \ldots$ For example, the elementary problem of determining how many different permutations of n different objects there are, can be reduced to discovering the appropriate function of n, which we could write as $p(n)$ or p_n for $n = 0, 1, 2, \ldots$ Other problems of the same general type involve two or more independent integer variables; to find the number of combinations of n things taken m at a time we seek a function which we can write as

$$c(m, n) \quad \text{or} \quad c_{m, n}, \quad {}^{n}C_{m} \quad \text{or} \quad \binom{n}{m}$$

depending on the notation chosen.

One way of approaching the desired solutions is by looking for a reasoned argument which will relate the unknown function for a general value n (in the one variable case) with values of the unknown function at one or more smaller values of n. If such a recurrence relation can be produced, it can usually be made the basis of an algorithm for computing values of the desired function, while in certain very useful classes of cases the relation can be solved to give an explicit expression for the unknown function. This chapter and the next are concerned with some of the techniques that are available for finding the explicit solutions of certain recurrence relations and with illustrating the sorts of combinatorial problems which are amenable to this approach.

2.1.1 Differences and Definitions

If u is a function of one variable n, which takes integer values $0, 1, 2, \ldots,$ the difference of the function values at two consecutive arguments

$$u_{n+1} - u_n$$

is called the first (forward) difference of u_n; it is often written

$$\Delta u_n = u_{n+1} - u_n$$

where Δ is the forward difference operator. An alternative way of writing the difference is in terms of the shift operator, E, where

$$E u_n = u_{n+1}$$

so that

$$\Delta u_n = E u_n - u_n$$

The operators Δ and E both require a certain operation to be performed on the functions to which they are applied, and, of course, are not ordinary algebraic quantities. In spite of this it will be found that much algebraic manipulation of Δ and E turns out to lead to legitimate results, and with this cautionary word we shall proceed to 'multiply' and to 'expand' them in what follows.

For example, since

$$\Delta u_n = (E-1) u_n$$

we have a formal relation

$$\Delta \equiv E-1 \tag{2.1}$$

The second difference (i. e. the difference of the first difference)

$$\Delta^2 u_n = \Delta u_{n+1} - \Delta u_n$$
$$= (u_{n+2} - u_{n+1}) - (u_{n+1} - u_n)$$
$$= u_{n+2} - 2u_{n+1} + u_n$$

can be derived in terms of the function values more swiftly by using (2.1);

$$\Delta^2 u_n = (E-1)^2 u_n$$
$$= (E^2 - 2E + 1) u_n$$

More generally for any positive integer k,

$$\Delta^k u_n = u_{n+k} - \binom{k}{1} u_{n+k-1} + \binom{k}{2} u_{n+k-2} - \ldots + (-1)^k u_n \qquad (2.2)$$

The relation can be used in the other sense, too. From (2.1)

$$E = 1 + \Delta \qquad (2.3)$$

and the function values are given in terms of differences by

$$u_{n+k} = E^k u_n = \Delta^k u_n + \binom{k}{1} \Delta^{k-1} u_n + \binom{k}{2} \Delta^{k-2} u_n + \ldots + u_n \qquad (2.4)$$

If there exists some relation $F(u_n, u_{n+1}, \ldots, u_{n+r}, n) = 0$ between the values of the unknown function u at a set of values of the independent variable n which has extent exactly $r + 1$, then the function u is said to satisfy a difference equation of order r. In the form quoted, the function values themselves appear and the equation is often called a recurrence relation. When it is equivalently expressed in terms of differences, say, in the form $g(u_n, \Delta u_n, \Delta^2 u_n, \ldots, \Delta^r u_n, n) = 0$, it is usually called a difference equation of order r.

If the relation between the function values and the differences is such that it is linear in the unknown function, so that no powers of u or products of u values at different arguments appear, we have a linear difference equation. Unless a non-linear equation happens to have a rather convenient form, its solution in explicit terms is likely to be impossible. A few examples where some progress can be made do occur, but most of our attention is directed towards the solution in explicit terms of linear difference equations, commencing with those in which all the coefficients of the unknown function or its differences are constants, i. e. the 'linear difference equation with constant coefficients'. The remainder of this chapter deals only with equations of this type.

2.2 THE HOMOGENEOUS EQUATION

The constant coefficient linear difference equation of order r can be written

$$a_o u_{n+r} + a_1 u_{n+r-1} + \ldots + a_r u_n = \phi(n) \qquad (2.5)$$

where the coefficients a_i $(i = 0, \ldots, r)$ are constants and the function $\phi(n)$ is given. The equation is in <u>homogeneous</u> form if $\phi(n) \equiv 0$. Clearly a solution of the <u>complete</u> (i. e. non-homogeneous) equation can have added to it any multiple of a solution of the homogeneous equation and the sum will still be a solution. Accordingly we now seek the general solution of (2.5) for the case $\phi(n) \equiv 0$ and in the next section we deal with more general cases.

By trial, it is evident that $u_n = m^n$ is a solution for those values of m which satisfy the equation

$$m^n \{ a_o m^r + a_1 m^{r-1} + \ldots + a_r \} = 0 \qquad (2.6)$$

The solution $m = 0$ is of no interest. The remaining condition on the value of m is called the <u>indicial equation</u>. This r^{th} degree polynomial equation has, of course, r roots and the form of the solution of the homogeneous difference equation depends upon whether the roots of the indicial equation are distinct or whether certain of them are repeated.

Case 1. Distinct Roots

Suppose the indicial equation of (2.6) has r different roots; let them be $m_1, m_2, \ldots, m_r$; then the general solution of the homogeneous equation is

$$u_n = A_1 m_1^n + A_2 m_2^n + \ldots + A_r m_r^n \qquad (2.7)$$

where $A_1,\ldots,A_r$ are constants which may be chosen to satisfy any boundary conditions upon the solution. The most commonly encountered such conditions are when the first r values $u_0, u_1, \ldots, u_{r-1}$ are known.

Example 1. To solve $u_{n+2} - 7u_{n+1} + 12u_n = 0$ subject to the boundary conditions $u_0 = 2, u_1 = 7$.
The indicial equation is

$$m^2 - 7m + 12 = 0$$

which has roots $m = 3, 4$, and so the general solution of the difference equation is

$$u_n = A3^n + B4^n$$

where A, B are any constants. In order to satisfy the boundary conditions, however, we must obtain the given values for u_0, u_1. Hence

$$A + B = 2$$
$$3A + 4B = 7$$

which give $A = B = 1$, and the required solution is

$$u_n = 3^n + 4^n$$

Case 2. Multiple Roots
   Where two or more roots of the indicial equation are coincident, the previous form of the solution (2.7) loses one or more of its independent constants A_i and it is clear that they could not be uniquely determined by r initial values of the u's. In this case we can try a solution of the form $u_n = m^n v_n$ and hope to obtain an equation for v_n that we can solve. In operator form the homogeneous equation (2.5) is

$$(a_0 m^{n+r} E^{n+r} + \ldots + a_r m^n E^n) v_0 = 0 \qquad (2.8)$$

which can be written as

$$m^n E^n [f(mE)] v_0 = 0 \tag{2.9}$$

where $f(x)$ is the polynomial in x of the indicial equation (2.6).

If now we write $E = 1 + \Delta$ and expand the polynomial f by Taylor's theorem, we can write (2.9) as

$$[f(m) + \frac{mf'(m)\Delta}{1!} + \frac{m^2 f''(m)\Delta^2}{2!} + \ldots + \frac{m^r f^{(r)}(m)\Delta^r}{r!}] v_0 = 0 \tag{2.10}$$

It was mentioned in the last section that we intended to 'expand' the operator expressions in cavalier fashion but that the results would justify the means. They do so here.

We note that the expansion terminates and that it has just $r + 1$ terms since f is a polynomial of degree r ; moreover since we have supposed that m is a multiple root of multiplicity k say, the first k-1 derivatives at $x = m$ must vanish.

$$f(m) = f'(m) = \ldots = f^{(k-1)}(m) = 0$$

Hence one solution of (2.8) is given by $\Delta^k v_0 = 0$. In just the same way that differentiating a polynomial of degree r produces a polynomial of degree $r - 1$, so differencing a polynomial reduces the degree by one, for

$$\Delta n^k = (n+1)^k - n^k$$

$$= kn^{k-1} + \text{terms with lower powers of } n$$

Accordingly the solution we need is

$$u_n = (A_1 + A_2 n + \ldots + A_k n^{k-1}) m_1^n + A_{k+1} m_{k+1}^n + \ldots + A_r m_r^n \tag{2.11}$$

Thus the part of the solution of a homogeneous equation corresponding to a root of the indicial equation which is repeated k times is the product of an arbitrary polynomial of degree $(k-1)$ in n and the root raised to the nth power.

Example 2 To solve $u_{n+3} - 3u_{n+1} - 2u_n = 0$.

The indicial equation is

$$m^3 - 3m - 2 = 0$$

This factorises to give $(m+1)^2(m-2) = 0$ which shows that the roots of the indicial equation are -1 (twice), 2. Accordingly the general solution is

$$u_n = (A_1 + A_2 n)(-1)^n + A_3 2^n$$

Example 3 To solve $u_{n+6} - 2u_{n+3} + u_n = 0$.

The indicial equation reduces to $(m^3 - 1)^2 = 0$, giving roots equal to the cube roots of unity, i.e. $m = 1$, $\exp(2i\pi/3)$, $\exp(-2i\pi/3)$ all twice repeated; the existence of complex roots of the indicial equation poses no problem although there is a little more algebra. One form of the general solution is

$$u_n = A_1 + A_2 n + (A_3 + A_4 n)\exp(2i\pi n/3) + (A_5 + A_6 n)\exp(-2i\pi n/3)$$

This may be expressed in a more convenient form by combining appropriate parts of the complex quantities (since $\exp(i\theta) = \cos\theta + i\sin\theta$)

$$u_n = A_1 + A_2 n + (B_3 + B_4 n)\cos 2\pi n/3 + (B_5 + B_6 n)\sin 2\pi n/3$$

where B_3, B_4, B_5, B_6 are also arbitrary constants related, of course, to $A_3, \ldots, A_6$ in the other form of the solution.

It will be noticed that the general solution of a linear difference equation with constant coefficients can have oscillatory components of constant, exponentially increasing or decreasing amplitude as well as the sums of powers of real constants and polynomials.

2.3 THE COMPLETE EQUATION

In the previous section the different cases of the homogeneous equation have been studied and solutions have been derived which incorporate the same number of arbitrary constants as the number describing the order of the equation; thus a second order equation has two arbitrary constants which can be determined from the boundary conditions which might, for example, give the values of the first two members of the sequence $\{u_n\}$. Since the complete equation (2.5) is linear, a general solution of the homogeneous equation can be added to any <u>particular solution</u> which can be found for the complete equation. The task of solving the complete equation therefore falls into two parts: first, to solve the homogeneous equation in full generality, and second, to discover any solution at all of the complete equation itself. The sum of these two parts provides the general solution to the complete equation and if appropriate boundary conditions are given, the arbitrary constants in the solution may be determined.

As is so often the case with mathematical problems, a good method is to guess the answer, or at least the form of the answer, and subsequently to verify it, identifying the coefficients in the guessed expression. Such an approach usually depends upon a little insight and more experience. In order to aid the one and provide the other, a more routine but slightly longer approach is presented first and then when the forms of the solutions to certain kinds of equations have been found, the method of trial solutions is exhibited.

2.3.1 A Symbolic Approach for Polynomials

Equations with certain types of known function $\phi(n)$ are easier to solve than others. One case that is straightforward is where $\phi(n)$ is a polynomial in n, say $P(n)$. The equation (2.5) can then be written using the symbolic operator E

$$(a_0 E^r + a_1 E^{r-1} + \ldots + a_r) u_n = P(n)$$

or if

$$f(E) = a_0 E^r + a_1 E^{r-1} + \ldots + a_r$$

symbolically we have a solution of the difference equation as

$$u_n = \frac{1}{f(E)} \cdot P(n) \qquad (2.12)$$

In the above, $f(E)$ is the polynomial in the operator E which specifies

12

the left hand side of the difference equation, so that finding a particular solution reduces to finding a way of interpreting the symbolic expression in (2.12). Since $P(n)$ is a polynomial, a plausible meaning seems in prospect if the E operator is replaced by the appropriate forward difference operator Δ. Accordingly, we substitute $1 + \Delta$ for E and expand the operator expression in powers of Δ. Thus

$$u_n = \frac{1}{f(1+\Delta)} \cdot P(n) = (a_0' + a_1'\Delta + \ldots + a_k'\Delta^k + \ldots)P(n) \tag{2.13}$$

where the a_i' $(i = 0, 1, 2, \ldots)$ are the constants in the expansion.

In ordinary circumstances one has qualms about expansions into infinite series lest they should be divergent, and when the 'variables' are in fact operators even the usual tests for convergence are not available to banish the worries. In this case, however, the infinite series in the operator Δ is being applied to a polynomial, so that if the terms in the expansion can be interpreted as they are written (an assumption certainly not proved here) only a finite number of non-zero terms will appear in the series since $\Delta^s P(n) = 0$ if s is higher than the degree of the polynomial $P(n)$. The interpretation is then a matter of just evaluating the differences of $P(n)$ and gathering the terms together. The mechanics of this step can sometimes be made easier if $P(n)$ is expressed in terms of the descending factorials $n^{(1)}, n^{(2)}, n^{(3)}, \ldots, n^{(s)}$ where $n^{(s)} = n(n-1) \ldots (n-s+1)$ since

$$\begin{aligned} \Delta n^{(s)} &= (n+1)^{(s)} - n^{(s)} \\ &= [n+1 - (n-s+1)]n^{(s-1)} \\ &= sn^{(s-1)} \end{aligned}$$

Thus for these discrete functions, the descending factorials, the differencing operator, Δ, behaves like the differentiation operator on the powers, since

$$\frac{d}{dx} x^n = nx^{n-1}$$

The proof that the above apparently unjustified manipulation of E's and Δ's is legitimate in these circumstances is omitted, but the results obtained can be readily tested by substitution in the original equation.

Example. Solve

$$u_{n+2} + u_{n+1} + u_n = n^2 + n + 1$$

The complementary function is found (as in section 2.2) by solving the indicial equation $m^2 + m + 1 = 0$; if its roots are m_1, m_2 , the complementary function is $A_1 m_1^n + A_2 m_2^n$ for arbitrary constants A_1, A_2 .

$$u_n = \frac{1}{E^2 + E + 1} \cdot (n^2 + n + 1)$$

This must now be written in terms of the Δ operator by replacing E by $1 + \Delta$,

$$u_n = \frac{1}{3 + 3\Delta + \Delta^2} \cdot (n^2 + n + 1)$$

$$= \frac{1}{3} (1 + \Delta + \frac{1}{3} \Delta^2)^{-1} \cdot (n^2 + n + 1)$$

In this form the polynomial term in Δ can be expanded by the binomial theorem, remembering that one need only include terms up to Δ^2 since the polynomial function being operated upon is quadratic. Hence

$$u_n = \frac{1}{3} [1 - \Delta - \frac{1}{3} \Delta^2 + (\Delta + \frac{1}{3} \Delta^2)^2 + \ldots] \cdot (n^2 + n + 1)$$

At this stage the powers in the polynomial in n should be replaced by the descending factorials. Hence

$$u_n = \frac{1}{3} [1 - \Delta + \frac{2}{3} \Delta^2 + \ldots] \cdot (n^{(2)} + 2n^{(1)} + 1)$$

since

$$n^{(2)} = n(n-1) \text{ and } n = n^{(1)}$$

The differencing of the polynomial can be more conveniently performed in terms of the descending factorials. The three terms of the operator now give

14

$$u_n = \frac{1}{3}[(n^{(2)} + 2n^{(1)} + 1) - (2n^{(1)} + 2) + \frac{4}{3}]$$

$$= \frac{1}{3}(n^2 - n + \frac{1}{3})$$

This expression is a particular solution of the equation given. The general solution is accordingly

$$u_n = A_1 m_1^n + A_2 m_2^n + \frac{1}{3}(n^2 - n + \frac{1}{3})$$

where m_1 and m_2 are the roots of the indicial equation $m^2 + m + 1 = 0$.

For certain equations $f(1+\Delta)$ will not have a non-zero constant term and in these cases the expansion of $[f(1+\Delta)]^{-1}$ into an infinite power series in Δ will not be possible. This will occur if $m = 1$ is a root of the indicial equation; $f(1+\Delta)$ will be found to have one or more factors, Δ, which must be removed before the expansion in (2.13) is attempted. After the interpretation of the series in Δ operating on the polynomial has been completed, one or more summing operations (as the inverse to the differencing operation) will be necessary.

Example. Find a particular solution of the equation

$$u_{n+4} - 5u_{n+3} + 9u_{n+2} - 7u_{n+1} + 2u_n = n^3 + 1$$

Here

$$f(E) \equiv (E-1)^3 (E-2) = \Delta^3 (\Delta-1)$$

Hence we have to interpret

$$\frac{1}{\Delta^3(\Delta-1)} \cdot (n^3+1) = \frac{-1}{\Delta^3} (1+\Delta+\Delta^2+\Delta^3+\ldots) \cdot (n^3+1)$$

All of the subsequent differencing and summing operations will be made easier if $n^3 + 1$ is written as $n^{(3)} + 3n^{(2)} + n^{(1)} + 1$, where, as before, $n^{(3)} = n(n-1)(n-2)$. Hence

$$u_n = -\frac{1}{\Delta^3}[n^{(3)} + 3n^{(2)} + n^{(1)} + 1 + 3n^{(2)} + 6n^{(1)} + 1 + 3.\,2n^{(1)} + 3.\,2 + 3.\,2.\,1]$$

15

$$= -\frac{1}{\Delta^3}[n^{(3)}+6n^{(2)}+13n^{(1)}+14]$$

The inverse operation to Δ on the descending factorials behaves similarly to the integration of powers, and so a particular solution is

$$u_n = -[\frac{1}{6.\,5.\,4}\,n^{(6)}+\frac{6}{5.\,4.\,3}\,n^{(5)}+\frac{13}{4.\,3.\,2}\,n^{(4)}+\frac{14}{3.\,2.\,1}\,n^{(3)}]$$

Since the indicial equation has the triply repeated root $m = 1$ together with the root $m = 2$, the complementary function contains an arbitrary quadratic in n which corresponds to the arbitrary lower degree terms which would appear from the repeated summations (just as arbitrary constants occur in integration). It is customary to ignore those terms in the particular solution and allow them to enter through the complementary function. The solution can, of course, be rewritten in terms of powers of n if required.

2.3.2 A Trial Solution for Polynomials

The previous section shows the form of the particular solution that is to be expected when the known function on the right hand side of the equation is a polynomial of degree k. If the indicial equation does not have any roots equal to unity, the particular solution will be a polynomial of degree k. If the indicial equation has a unit root repeated j times, the particular solution is a polynomial of degree $j+k$. The method of trial solutions first examines the indicial equation and then assumes for the solution a general polynomial of the appropriate degree.

Example. Solve

$$u_{n+2} + u_{n+1} + u_n = n^2 + n + 1$$

The indicial equation

$$m^2 + m + 1 = 0$$

does not have $m = 1$ as a root. Hence the particular solution will be a quadratic in n : accordingly we assume a trial solution

16

$$u_n = an^2 + bn + c$$

When this form is substituted in the equation we obtain

$$a(n+2)^2 + b(n+2) + c + a(n+1)^2 + b(n+1) + c + an^2 + bn + c = n^2 + n + 1$$

Since a particular solution must satisfy the equation identically we may equate coefficients of the powers of n and so obtain simultaneous linear equations for the unknown coefficients a, b, c

$$3a = 1$$
$$6a + 3b = 1$$
$$5a + 3b + 3c = 1$$

The solutions of these equations yield the particular solution $u_n = \frac{1}{3}(n^2 - n + \frac{1}{3})$ which was found in the first example in section 2.3.1.

In the other example of that section, an appropriate expression for the form of the particular solution can only be obtained by noting that the indicial equation has the root $m = 1$ repeated three times so that a polynomial of 6th degree needs to be tried. It can be seen at once that more than a line or two of algebra will be involved in this approach and that there are many possibilities for arithmetic error. The task can be lightened a little by casting the equation in Δ form and assuming a trial solution expressed in terms of the descending factorials.

Thus, the example

$$u_{n+4} - 5u_{n+3} + 9u_{n+2} - 7u_{n+1} + 2u_n = n^3 + 1$$

is equivalent to

$$\Delta^4 u_n - \Delta^3 u_n = n^{(3)} + 3n^{(2)} + n^{(1)} + 1$$

The trial solution is best written in the form

$$u_n = an^{(6)} + bn^{(5)} + cn^{(4)} + dn^{(3)}$$

Note that we do not need to include terms in $n^{(2)}, n^{(1)}$ or a constant term since these will appear in the complementary function. The differencing of the trial solution in this form is easy, giving coefficients of the descending factorials which can also be equated in order to obtain an identity and so to produce the equations

$$
\begin{aligned}
-120a &= 1 \\
360a - 60b &= 3 \\
120b - 24c &= 1 \\
24c - 6d &= 1
\end{aligned}
$$

which again recovers the previous solution.

2.3.3 Symbolic Solutions for Exponentials

Another equation for which the particular solution may be easily found has the known function as some multiple of a^n where a is a constant. The simplicity occurs because the shift operator, E, has the same effect on the function a^n as the straightforward multiplication by a; $Ea^n = a^{n+1} = a.a^n$. More generally $E^r.a^n = a^r.a^n$ so that $g(E).a^n = g(a)a^n$ where $g(E)$ is any polynomial function in the operator E. Thus the difference equation

$$
f(E)u_n = a^n \tag{2.14}
$$

will be found to have particular solution

$$
u_n = a^n/f(a) \tag{2.15}
$$

as long as $f(a) \neq 0$.

If the indicial equation has a root $m = a$ so that both $f(a) = 0$ and the complementary function includes an arbitrary multiple of a^n itself, the above argument needs some modification. In such a case the polynomial $f(x)$ will have one or more factors $(x-a)$. Let $f(E) \equiv (E-a)^k g(E)$ where $g(E)$ is a polynomial in E and $g(a) \neq 0$. Then

$$
(E-a)^k . u_n = \frac{1}{g(E)} . a^n = \frac{a^n}{g(a)} \qquad \text{as before.}
$$

18

In order to make further progress it is now necessary to write u_n as a product of a known function of n and an unknown one, in a manner that will appear yet again in this chapter and in a later one (when considering a reduction of the order of the difference equations).

In this case we put

$$u_n = a^n v_n$$

Since $E^r(a^n v_n) = a^{n+r} v_{n+r} = a^n[(aE)^r]v_n$,

$$(E-a)^k u_n = a^n(aE-a)^k v_n$$

so that we have the equation

$$a^k(E-1)^k v_n = \frac{1}{g(a)}$$

which is in the form treated in the latter part of section 2.3.1. Its solution therefore is

$$v_n = \frac{1}{a^k g(a)} \ \frac{1}{\Delta^k} \cdot 1$$

A particular solution for v_n excluding the terms with arbitrary constants is

$$v_n = \frac{n^{(k)}}{a^k g(a)k!} \qquad (2.16)$$

so that

$$u_n = \frac{a^{n-k}}{g(a)} \ \frac{n^{(k)}}{k!} \qquad (2.17)$$

It is perhaps not surprising to find this form of particular solution; it is a natural generalisation of the corresponding one (a polynomial) when the indicial equation has a root of unity.

2.3.4 A Trial Solution for Exponentials

A particular solution corresponding to a term on the right hand side, proportional to a^n, say, has been shown in the last section to be itself proportional to a^n as long as a is not a root of the indicial equation. If it is a root then the particular solution is a^n multiplied by a polynomial in n of degree equal to the multiplicity of the root a of the indicial equation.

Example. Solve

$$u_{n+2} - u_n = 3.2^n + 4(-1)^n$$

As usual, we first look at the indicial equation $m^2 - 1 = 0$, which has roots ± 1. The particular solution is found in two parts, considering the terms on the right hand side separately. Thus for the term 3.2^n we assume a particular solution of the form $k.2^n$ and immediately find that $k = 1$. Alternatively we could use the operator method

$$\frac{1}{E^2 - 1} \cdot 3.2^n = \frac{3.2^n}{2^2 - 1} = 2^n$$

For the other term, $4(-1)^n$, we know that a particular solution will be a linear function of n multiplied by $(-1)^n$. Thus we can assume

$$u_n = an(-1)^n$$

so that substitution in the equation gives

$$a(n+2)(-1)^{n+2} - an(-1)^n = 4(-1)^n$$

or $2a = 4$. The solution is accordingly

$$u_n = A + (B+2n)(-1)^n + 2^n$$

where A and B are the constants of the complementary function.

It is worth noting that the exponential case also includes those cases involving certain of the trigonometrical functions. We can treat, for example, cos kn as the real part of e^{ikn}, solve the equation for this exponential, and merely identify the real part of the solution.

Clearly the same analysis produces a solution to the difference equation which has the same right hand side apart from sin kn replacing the cos kn; in this case we must take instead imaginary parts of the solution for e^{ikn}. More complicated combinations of sines and cosines can be dealt with by expressing them in terms of multiple angles by the usual formulae.

Example. To solve

$$u_{n+2} + a^2 u_n = \cos kn$$

it is convenient to consider the equation

$$v_{n+2} + a^2 v_n = e^{ikn}$$

A particular solution of this equation is

$$v_n = \frac{1}{E^2 + a^2} \cdot e^{ikn}$$

$$= \frac{e^{ikn}}{a^2 + e^{2ik}}$$

A particular solution of the original equation is the real part of this expression - which can be obtained if numerator and denominator are multiplied by $a^2 + e^{-2ik}$; using $e^{2ik} + e^{-2ik} = 2\cos 2k$ we have

$$u_n = \frac{a^2 \cos kn + \cos(k-2)n}{a^4 + 2a^2 \cos 2k + 1}$$

Rules for the trial solution could of course be adopted instead of the operator approach once the equation has an exponential appearing on the right hand side, but it requires rather more experience of solving equations with sines and cosines to guess the form of the solution if the device of treating the known function as the real (imaginary) part of a complex exponential is not adopted.

2.3.5 More General Known Functions

A particular solution to the constant coefficient linear difference equation with a general right hand side

$$f(E). u_n = \phi(n) \tag{2.18}$$

is given in purely formal terms by

$$u_n = \frac{1}{f(E)} \cdot \phi(n) \tag{2.19}$$

but the interpretation of the operator function on $\phi(n)$ needs some manipulation. Since $f(E)$ is a polynomial, a partial fraction expansion is available for $\frac{1}{f(E)}$, say

$$\frac{1}{f(E)} \equiv \frac{A_1}{(E-m_1)} + \frac{A_2}{(E-m_1)^2} + \ldots + \frac{A_k}{(E-m_1)^k} + \frac{A_{k+1}}{(E-m_2)} + \ldots$$

where, for example, m_1 is a root of the indicial equation which is repeated k times. In order to obtain a particular solution we treat each term on the right separately and so we need to solve equations like

$$v_n = \frac{1}{(E-m)^s} \cdot \phi(n) \tag{2.20}$$

or equivalently, to find v_n such that

$$(E-m)^s v_n = \phi(n)$$

Again we make the substitution $v_n = m^n w_n$. Just as in section 2.3.3,

$$(E-m)(m^n w_n) = m^{n+1}(E-1)w_n$$

and more generally

$$(E-m)^r(m^n w_n) = m^{n+r}(E-1)^r w_n$$

Hence, a solution to the equation (2.20) is

22

$$w_n = \frac{1}{\Delta^r} \cdot \{m^{-n-r} \phi(n)\}$$

and hence the appropriate part of the particular solution of the equation (2.18) is

$$v_n = m^n \frac{1}{\Delta^r} \cdot \{m^{-n-r}\phi(n)\} \tag{2.21}$$

The particular solution of the equation thus depends upon a sum of a number of terms each of which may itself depend upon a single or repeated summation of quantities involving the known right hand side of the equation. Even though the repeated summations may be expressed in terms of a single summation by a formula corresponding to that of Cauchy for repeated integration (see, e.g., Milne-Thomson section 8.12) it will only be in special cases that the result can be expressed explicitly in terms of elementary functions.

2.4   SOLUTION BY GENERATING FUNCTIONS

A recurrence relation or difference equation states a relation between members of a sequence, say $\{u_n\}$ $(n = 0, 1, 2, \dots)$; usually the most convenient way of describing such a sequence is to give an explicit formula for its general member u_n but another way is to quote a function of a dummy variable, x (the function being called a <u>generating function</u> or G.F.) which has an unique expansion in a convergent series of powers of x (or of another convenient set of linearly independent functions of x) whose coefficients are the quantities $u_0, u_1, u_2, \dots$ For example, the sequence $\{n\}$ can be described using the powers of x by a generating function $F(x)$ where

$$F(x) = \sum_{n=0}^{\infty} n x^n$$

$$= x \cdot \frac{d}{dx} \left[\sum_{i=1}^{\infty} x^n \right]$$

$$= \frac{x}{(1-x)^2}$$

If this function were given, it can, of course, be expanded in powers of x and so yield any desired coefficient.

In order to be useful, a generating function needs to converge for a suitable range of the dummy variable, so that if the sequence $\{u_n\}$ increases rapidly the powers of x may not be suitable for the construction of a generating function. In such a case another set of linearly independent functions like $\{x^n/n!\}$ may serve. For example the sequence $\{u_n\}$, where $u_n = n!$, can be described by a generating function $(1-x)^{-1}$ since

$$\sum_{n=0}^{\infty} u_n \cdot \frac{x^n}{n!} = \sum_{n=0}^{\infty} x^n = (1 - x)^{-1}$$

Thus generating functions are convenient ways of describing an infinite sequence of numbers but they can be used in many ways to obtain properties of the sequence. Their main importance for us in this part of the subject is that they can be an aid in obtaining an explicit expression for the solution, u_n , of some difference equations.

Clearly they will not be greatly used in solving linear constant coefficient difference equations because methods previously described are quite simple, although there are cases in which some rather awkward particular solutions may be obtainable more easily via a G. F. However these equations allow the method to be introduced with few complications.

For example, the G. F. approach to solve

$$u_{n+1} - 2u_n = na^n$$

considers the infinite set of equations

$$u_1 - 2u_0 = 0$$
$$u_2 - 2u_1 = a$$
$$u_3 - 2u_2 = 2a^2$$
$$- - - - - - - -$$
$$- - - - - - - -$$

In order to derive the G. F. $G(x) = \sum_{n=0}^{\infty} u_n x^n$, the nth equation of the set is multiplied by x^n and the set is summed. We obtain

$$\sum_{n=1}^{\infty} u_n x^n - \sum_{n=0}^{\infty} u_n x^{n+1} = \sum_{n=0}^{\infty} na^n x^{n+1}$$

or

$$(G(x)-u_0) - 2xG(x) = ax^2/(1-ax)^2$$

Hence

$$G(x) = \frac{u_0}{1-2x} + \frac{ax^2}{(1-2x)(1-ax)^2}$$

which is the G. F. of the sequence $\{u_n\}$.

To find the explicit expression for u_n we need to expand the G. F. - by expressing it in partial fractions

$$G(x) = \frac{u_0}{1-2x} + \frac{1}{(a-2)(1-ax)^2} - \frac{2(a-1)}{(a-2)^2(1-ax)} + \frac{a}{(a-2)^2(1-2x)}$$

Each term can then be expanded in powers of x. The first and the last terms composing $G(x)$ have the factor $(1-2x)^{-1}$, which corresponds to the complementary function, A. 2^n, and they show how the boundary value, u_0, enters the solution. The selection of the coefficient of x^n in the expansion of the other terms gives a particular solution

$$u_n = \frac{na^n}{(a-2)} - \frac{a^{n+1}}{(a-2)^2}$$

The solution of a general linear equation with constant coefficients is obtained in a similar way but at the expense of even more algebra, so that generating functions are rarely employed in this case; in certain equations which we meet later the generating function approach is one of the most convenient.

2.5 SIMULTANEOUS EQUATIONS

The members of two or more sequences may be related by a set of simultaneous linear difference equations with constant coefficients; just as linear algebraic simultaneous equations are solved by eliminating the unknowns systematically, so we can proceed in the corresponding manner remembering that the elimination will require differencing or shifting operations instead of simple multiplications.

Example. To solve the pair of simultaneous difference equations

$$u_{n+1} - u_n + v_n = n$$

$$u_n + 2v_{n+1} - v_n = 2$$

they should be written in operator form

$$(E-1)u_n + v_n = n \tag{2.22}$$

$$u_n + (2E-1)v_n = 2 \tag{2.23}$$

The variables u_n can be eliminated by operating on (2.23) with the operator $(E-1)$; equation (2.23) becomes

$$(E-1)u_n + (E-1)(2E-1)v_n = (E-1)2 \tag{2.24}$$

Since shifting n to $n+1$ in a constant has no effect, $E.2 = 2$; subtracting (2.22) from equation (2.24) yields

$$(2E^2 - 3E)v_n = -n$$

which is a (<u>first</u> order) equation with solution

$$v_n = (n+1) + A\left(\frac{3}{2}\right)^n$$

This solution for v_n can be substituted in either of the original equations - conveniently the second one in this case, which is then solved for u_n , giving

$$u_n = -(n+1) - 2A\left(\frac{3}{2}\right)^n$$

2.6 APPLICATIONS

2.6.1 Algol Expressions

Clearly a great many different Algol expressions can be formed from the permitted symbols. In order to illustrate a method, we consider how many valid Algol expressions can be formed using exactly k symbols drawn from the set $\{0, 1, 2, 3, 4, 5, 6, 7, 8, 9, +, -, /, \times\}$, each with whatever repetition is desired.

The number of expressions that we have to find clearly depends upon the syntax of Algol and the set of symbols we are allowed to use, but more significantly for our start upon the problem, it depends upon the number, k, of symbols in the expression. It is reasonable, therefore, for us to let the number we require be u_k and to attempt to derive a relationship between u_k and $u_{k-1}, u_{k-2}, \ldots$ In order to do so we must now use a knowledge of the syntax and the symbols available. An expression cannot end with an arithmetic sign and so all the valid arrangements of the symbols must have a digit at the end. The last symbol but one must therefore either be a digit, in which case the first k-1 symbols form a valid Algol expression, or that symbol must be one of the four arithmetic signs in which case the first k-2 symbols must form a valid Algol expression. In BNF notation this part of the syntax can be written

$$< \text{expression}> :: = < \text{expression}> < \text{digit}> \; | < \text{expression}> < \text{sign}> < \text{digit}>$$

Thus the expressions using just k symbols taken from the set of those permitted can be divided into two mutually exclusive and exhaustive classes, namely those with a digit in the penultimate position and those with a sign there. The number in the first class must be $10u_{k-1}$ since there are ten different digits which may be appended to a valid expression of k-1 symbols. The second class has $40u_{k-2}$ since each one of the four signs may be followed by any one of the ten digits and both appended to a valid expression of k-2 symbols. Hence we have the recurrence relation

$$u_k = 10u_{k-1} + 40u_{k-2}$$

This second order linear difference equation, valid for $k \geq 2$, needs two boundary conditions for its solution. The number of single symbol expressions is just the number of different digits, hence $u_1 = 10$. Similarly the number of two symbol expressions is 120, being formed of the 100 pairs of digits $(00, 01, \ldots, 99)$ and 20 others formed by a digit following one of the signs + or - .

The equation can be solved in the usual manner by obtaining the roots of the indicial equation

$$m^2 - 10m - 40 = 0$$

If the roots are α, β so that $\alpha, \beta = 5 \pm \sqrt{65}$, we have

$$u_k = A\alpha^k + B\beta^k$$

where

$$10 = A\alpha + B\beta$$
$$120 = A\alpha^2 + B\beta^2$$

from the initial conditions. The solution of these simultaneous linear equations gives

$$A = \frac{120 - 10\beta}{\alpha(\alpha - \beta)} \, , \quad B = \frac{120 - 10\alpha}{\beta(\beta - \alpha)}$$

These expressions and that for u_k can of course be expressed in terms of the appropriate surds (i. e. the square root quantities), but in this and in similar problems it is wise to defer the manipulation of the surds until as late as possible in order to reduce algebra - in some cases that labour and the unattractive complications of the algebra may be avoided.

A part of every example where special care is needed is in the separation of the items we are counting into two mutually exclusive but exhaustive classes, or in plainer terms to make sure that we are not counting some items twice and that all items are indeed being counted. The key idea in this example is that 'every valid expression of k symbols must either have its penultimate symbol a sign or a digit'. It clearly must have one or the other and it cannot have both, therefore it is legitimate for us to count these occurrences separately and then add them together. Problems are more or less difficult depending on how difficult it is to spot a way of dividing the class to be counted into such subclasses.

28

2.6.2 Colouring Sectors

A circular disk has been divided into a number of sectors, and there are given several different coloured paints with which we are required to paint the sectors on this disk in such a way that no adjacent sectors have the same colour paint upon them. We shall calculate the number of ways of painting an n sectored disk with p paints.

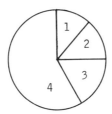

In this example, the number we are seeking clearly depends upon both n and p, but we would like if we can to avoid having to involve both of these variables in forming a recurrence relation; difference equations of more than one variable are more complicated to solve than single variable ones (compare partial differential equations and ordinary ones), so it is worth trying to see if a relation can be discovered which links the number we are seeking, either with the same problem for a smaller number of sectors, or for one with a smaller number of paints, but not both. Concentration on the number of sectors turns out to be a more fruitful approach. Let us therefore denote the number of ways of painting the disk by u_n and imagine sectors numbered as shown in the figure. Now we have the task of dividing all the painted n sectored disks into two mutually exclusive and exhaustive classes. We can do this by asserting that of these disks either sectors 1 and 3 have the same paint, or they have different paints on them. In the first case we could remove sector number 2 and imagine sectors 1 and 3 coalesced to leave a disk of n-2 sectors painted according to the rules. In this case the sector number 2 could have p-1 different paints upon it and so the total number of n sectored disks of this kind is $(p-1)u_{n-2}$. In the other case since sectors 1 and 3 are coloured differently if we remove sector 2 we are left with an n-1 sectored disk of a kind we are considering; since sector 2 can just have p-2 paints upon it, the total number of disks of this kind is $(p-2)u_{n-1}$. Hence we have the recurrence relation

$$u_n = (p-2)u_{n-1} + (p-1)u_{n-2}$$

Since this is an equation in the independent variable n , the coefficients are constants and it can be solved by the usual indicial equation approach. Two initial conditions are needed for a second order equation; although there is an argument for suggesting that $u_1 = p$, it seems slightly dangerous to use the case $n = 1$ as there is some ambiguity about whether the disk is considered as a whole without any sector boundary, or whether it is a sector with a left hand and right hand edge which happen to coincide - when it might be thought that the two adjacent edges would have the same colour either side of them. So for safety we take the boundary conditions for the cases $n = 2$ when $u_2 = p(p-1)$ and for $n = 3$ when $u_3 = p(p-1)(p-2)$. The indicial equation is

$$m^2 - (p-2)m - (p-1) = 0$$

which has roots $p-1$ and -1 . The solution accordingly is

$$u_n = A(p-1)^n + B(-1)^n$$

The two boundary conditions lead to the number of ways of painting the disk

$$u_n = (p-1)\,[(p-1)^{n-1} + (-1)^n]$$

The result $u_1 = 0$ confirms the suspicion of ambiguity for the case $n = 1$ and the correctness of the slightly unusual second alternative interpretation described above.

2.7 BIBLIOGRAPHY

Many books on elementary algebra as well as more advanced volumes on the finite difference calculus contain the material described here. The classic volume, by now rather old fashioned in appearance but crystal clear in its explanations, is 'The Calculus of Finite Differences' by L. M. Milne-Thomson (Macmillan, 1951), in which chapter 13 gives a quite detailed account with many examples.

2.8 EXERCISES

Solve the following equations:

[2.1] $u_{n+2} + 4u_{n+1} - 12u_n = 0$

[2.2] $\Delta^2 u_n + \Delta u_n = 0$

[2.3] $u_{n+6} + 2u_{n+3} + u_n = 0$

[2.4] $u_{n+2} - u_{n+1} - 6u_n = n$

[2.5] $u_{n+2} - 5u_{n+1} + 6u_n = 5^n$

[2.6] $u_{n+2} + 4u_{n+1} + 4u_n = n$

[2.7] $u_{n+3} - 5u_{n+2} + 8u_{n+1} - 4u_n = n2^n$

[2.8] $u_{n+2} - 3u_{n+1} - 4u_n = m^n$

[2.9] $u_{n+2} + 2u_{n+1} + u_n = n(n-1)(n-2) + n(-1)^n$

[2.10] $u_{n+2} - 2mu_{n+1} + (m^2+p^2)u_n = m^n$

[2.11] $(\Delta^2 - \Delta)y_n = 2^n + 3^n$ where $y_0 = 0$, $y_1 = 1$

[2.12] $(\Delta^2 - 2\Delta)y_n = 2^n + 3^n$ where $y_0 = 0$, $y_1 = 1$

[2.13] Let P_k be the number of permutations of n letters taken k together (with repetition) but no three consecutive letters being the same. Derive a difference equation connecting P_k, P_{k-1}, P_{k-2} and hence find P_k.

[2.14] (a) Solve the difference equation

$$u_n = u_{n-2} + n - 2 \qquad\qquad \text{(A)}$$

(b) A linear triplet of integers (a, b, c) is such that $a < b < c$ and $b-a = c-b$. If u_n is the number of distinct linear triplets that may be formed from the integers $1, 2, \ldots, n$, show that

$$u_{2n+1} = u_{2n} + n$$

Derive a similar relation for u_{2n}. Hence show that u_n satisfies (A) and obtain an explicit formula for u_n.

(Newcastle 1972)

[2.15] (i) Solve the difference equations

$$u_{n+1} = 2u_n + 2v_n$$

$$v_{n+1} = u_n + 2v_n$$

(ii) Consider permutations of the integers $1, 2, \ldots, n$ such that no integer is more than 2 greater than both its neighbours (or than its single neighbour for the end values). For example, for six integers the permutation 1 2 4 6 3 5 is acceptable since $6 - 4 \not> 2$ even though $6 - 3 > 2$.

Let u_n be the number satisfying these conditions in which $n-1$ and n are in adjacent positions, and let v_n be the corresponding number of permutations in which $n-1$ and n are not adjacent. Prove that u_n and v_n satisfy the difference equations in (i). Hence find the total number of n-permutations satisfying the conditions in (ii).

[2.16] Find the number of permutations of $1, 2, \ldots, n$ such that no integer is more than one place removed from its position in the natural order $(1, 2, \ldots, n)$.

[2.17] Solve the equations

$$u_{n+1} - u_n + v_n = n$$

$$u_n + 2v_{n+1} - v_n = 2$$

by first eliminating the unknowns v_n, v_{n+1}.

[2.18] (i) If $au_n + bu_{n-1} + cu_{n-2} = 0$, derive the recurrence relation satisfied by alternate members of the sequence $\{u_n\}$, i.e. express u_{2n} in terms of u_{2n-2}, u_{2n-4}, ...
(ii) Solve the equations

$$u_{n+1} = 2u_n + v_n$$

$$v_{n+1} = u_n + v_n$$

where $u_1 = v_1 = 1$.
(iii) Hence, or otherwise, show that $u_n + v_n = f_{2n}$ where f_n is a Fibonacci number satisfying

$$f_n = f_{n-1} + f_{n-2}, \quad f_0 = f_1 = 1$$

(iv) At a given moment, a memory of n cells is allocated into a number of blocks of consecutive cells, possibly separated by unallocated cells. Two allocations are considered identical only if their blocks are of the same sizes and are in the same locations in memory. Formulate difference equations for u_n, the number of different allocations of n cells, which have the first cell allocated to some block, and for v_n, the corresponding number in which the first cell is free.

(Newcastle 1974)

[2.19] If $x_{n+2} = 2(x_{n+1} + x_n)$, $x_0 = x_1 = 1$, find the limit of x_{n+1}/x_n as $n \to \infty$.

[2.20] An ancient method of calculating $\sqrt{2}$ is equivalent to putting

$u_0 = v_0 = 1$, calculating iteratively

$$u_n = u_{n-1} + 2v_{n-1}$$

$$v_n = u_{n-1} + v_{n-1}$$

and noticing that $(u_n/v_n)^2$ approaches 2 as n increases. Prove this result and find what is the least value of n for which u_n/v_n gives $\sqrt{2}$ correct to 10 decimal places.

[2.21] Find the number of binary sequences (i.e. sequences of zeros and ones only) of length n which do not contain two consecutive zeros.

[2.22] Find the number of quaternary sequences (i.e. of zeros, ones, twos, and threes only) of length n which contain an even number of zeros and an even number of ones.

[2.23] Derive an expression for the number of binary sequences of length n which contain the pattern 010 for the first time at the end.

[2.24] Calculate A^{100} where A is the matrix $\begin{pmatrix} 3 & -1 \\ 0 & 2 \end{pmatrix}$.

(University of Linkoping)

[2.25] If $a_n = a_{n-1} + a_{n-2}$ (n = 2, 3, ...) and $a_0 = 0$, $a_1 = 1$ show, by solving this equation or otherwise, that $a_n^2 + a_{n+1}^2 = a_{2n+1}$.

(Newcastle 1977)

[2.26] (a) A sequence $\{u_i\}$ is defined by

$$u_0 = 0$$

$$u_{n+1} = 2u_n + 2^n \quad (n = 0, 1, ...)$$

Find a generating function for the sequence $\{u_i\}$ and an expression for u_n.

(b) Let the total number of ones in the list of integers $0, 1, ..., n-1$ in binary notation be $G(n)$.

34

(i) Show that $G(2^{m+1}) = 2G(2^m) + 2^m$, and hence (or otherwise) that $G(2^m) = m2^{m-1}$.

(ii) By considering separately the even members and odd members of the set $\{i\}$ $(0 < i < 2n)$ (or otherwise) prove that $G(2n) = n+2G(n)$.

(iii) Find the corresponding relation for $G(2n+1)$.

(iv) Using b(ii) and (iii) above, derive an equation for a generating function for $\{G(n)\}$. Why is this difficult to solve?

(c) Given n sets, each consisting initially of one member. At step i $(i = 1, 2, \ldots, n-1)$, merge any two sets. If the cost of merging two sets is equal to the size of the smaller set, show (by considering the final step) that the maximum cost, $H(n)$, that can be incurred in the whole process satisfies

$$H(1) = 0$$

$$H(n) = \max_{1 \le i \le \frac{n}{2}} [i + H(i) + H(n-i)] \quad (n = 2, 3, \ldots)$$

(Newcastle 1976)

3 · Other Difference Equations

3.1 VARIABLE COEFFICIENTS

The complementary functions in the solutions of linear equations with constant coefficients have been shown to be sums of terms of the form $P(n)m^n$ where $P(n)$ is a polynomial in n and m is constant. Examples in the last chapter have shown how this part of the solution can exhibit sinusoidal behaviour of varying amplitude as well, but the unvarying feature is that the solution can grow or diminish no faster than the power of some constant. Apart, therefore, for those constant coefficient linear equations in which the known function takes forms other than the simple ones we have considered, e.g. factorials, the solution of the complete equation can not increase or decrease more quickly than an exponential. Many combinatorial problems, such as those concerned with permutations, involve quantities which exhibit a rapid increase, and these can appear as the solutions of difference equations in which the coefficients are no longer constant. For example, the function $u_n = n!$ satisfies the simple first order recurrence relation $u_n = nu_{n-1}$ with $u_0 = 1$. The linear equation with general coefficients can hardly be expected to have an explicit solution in terms of known functions, but there are a number of cases where the solution can be exhibited in a reasonably compact form and in the next sections we consider some of these cases.

3.1.1 First Order Equations

The general first order equation can be brought to the form

$$u_{n+1} - p_n u_n = r_n \tag{3.1}$$

in which both p_n and r_n are supposed to be known functions of n. Consider first the homogeneous equation in which $r_n \equiv 0$. By repeatedly applying the recurrence relation for $n = 1, 2, 3, \ldots$ we have

$$u_n = p_1\, p_2 \cdots p_{n-1}\, u_1 \tag{3.2}$$

It is no longer possible to add a particular solution to a complementary function to obtain the solution to the complete equation in which $r_n \neq 0$ because of its variable coefficient, but it can be solved by the same device that has been used previously. Let u_n be the product of the now known solution of the homogeneous equation and an unknown function v_n, i.e. let

$$u_n = v_n \prod_{i=1}^{n-1} p_i \tag{3.3}$$

The equation then becomes

$$v_{n+1} \prod_{i=1}^{n} p_i - v_n \prod_{i=1}^{n} p_i = r_n \tag{3.4}$$

Both sides of the equation can be divided by the continued product to give

$$v_{n+1} - v_n = s_n$$

where $s_n = r_n / \prod_{i=1}^{n} p_i$. The solution of this equation is found by summation which introduces a constant C (related, of course, to the boundary condition)

$$v_n = \sum_{i=1}^{n-1} s_i + C$$

which gives the solution of the original equation (3.1) to be

$$u_n = C \prod_{i=1}^{n-1} p_i + \prod_{i=1}^{n-1} p_i \sum_{i=1}^{n-1} \left(r_i / \prod_{j=1}^{i} p_j \right) \tag{3.5}$$

The first order equation with variable coefficients can accordingly always be solved, at least in terms of these continued products and a summation of them; although it will not necessarily be possible to express the result in a very tidy form.

3.1.2 Polynomial Coefficients

One approach for second and higher order equations with polynomial coefficients is to use the difference equation to derive a relation for an appropriate generating function. In general the relation derived will be a differential equation, which is sometimes soluble easily and conveniently although increasing complexity in the original difference equation quite rapidly produces either an unmanageable differential equation or a solution which presents difficulties when one tries to expand it to obtain the coefficients of the series expansion.

3.1.3 Combinations

For an illustration of the generating function solution consider the first order equation

$$(n+1)u_{n+1} - (c-n)u_n = 0$$

First define the generating function $G(x) = \sum u_n x^n$. Then the equations for $n = 0, 1, 2, \ldots$ are multiplied respectively by $x^0, x^1, x^2, \ldots$ and summed to obtain

$$\sum_0^\infty (n+1)u_{n+1}x_n - \sum_0^\infty (c-n)u_n x^n = 0$$

It is now possible to identify a part of this relation in terms of $G(x)$ and its derivatives since $\dfrac{dG}{dx} = \sum_{n=1}^\infty nu_n x^{n-1}$. In some equations care is necessary in case one or more terms involving $u_0, u_1, \ldots$ need special treatment but in this case it follows at once that

$$\frac{dG}{dx} - cG + x \sum_0^\infty nx^{n-1}u_n = 0$$

or $\quad (1+x)\dfrac{dG}{dx} = cG$

This first order differential equation is of the 'variables separable' type so that we can write all the terms involving G on one side and those with x on the other:

$$\frac{dG}{G} = c\frac{dx}{1+x}$$

which leads to a solution by integrating both sides

$$\log G = c \, \log(1+x) + A$$

where A is an arbitrary constant of integration. Exponentiating both sides and writing $K = e^A$ gives the generating function

$$G(x) = K(1+x)^c$$

The solution of the original difference equation, u_n , is the coefficient of x^n in the expansion of $G(x)$ and in this case a familiar explicit solution appears:

$$u_n = K \binom{c}{n}$$

There are a number of points to note in this simple example. First, the appearance of a first order differential equation for the generating function, has nothing at all to do with the original difference equation being of first order itself; it is more concerned with the coefficients in the difference equation being first degree polynomials in n . Second, and more importantly perhaps, care has to be taken in the form assumed for the generating function. We have needed to assume that the term by term differentiation of the generating function is equal to the derivative of the generating function itself. This is indeed the case for uniformly convergent series and in particular for convergent power series, as we have in this case. For us to carry out these manipulations the series for the generating function assumed must converge in some non-vanishing region about the origin $x = 0$, and for this to be the case the coefficients in the power series u_n must not increase too rapidly. If it is suspected that they will, then we need to take a different form for the generating function, for example to let $G(x) = \sum u_n x^n / n!$ and to obtain a corresponding equation for it.

3.1.4 Successors in Permutations

A permutation of the first n natural numbers is said to contain r correct successors if there are r positions in which the element i is preceded immediately by $i-1$ for positive values of i . Thus the permutation $(2 \; 3 \; 1)$ of the first three integers contains one correct successor.

The maximum number of correct successors in a permutation of n is clearly $n-1$, while there can be permutations with no correct successors at all. The task is to find the number $u_{n,r}$ of permutations of n which contain exactly r correct successors for $r = 0, 1, \ldots, n-1$.

In this example it seems that we are inevitably headed towards a two variable recurrence relation for $u_{n,r}$ but some simplification can be made immediately by noticing that correct successors can only arise from r of the digits $2, 3, \ldots, n$ following their correct predecessors and these r digits can be chosen from the $(n-1)$ possible ones in $\binom{n-1}{r}$ ways. In order to have just these r correct successors there must be no correct successors in the other positions - which a little thought will show can be obtained in $u_{n-r,0}$ ways. It is as though r of the items are bound together with their predecessors and the resulting $n-r$ bundles of 1, 2 or more digits are arranged so that there are no correct successors in them. Accordingly

$$u_{n,r} = \binom{n-1}{r} u_{n-r,0}$$

This relation will allow us to express the two variable function $u_{n,r}$ in terms of the one variable function which gives the number of permutations of n which possess no correct successors. We can therefore concentrate upon deriving a relationship for $u_{n,0}$ and do so by the technique already displayed of dividing the permutations with no correct successors into two exclusive and exhaustive classes. First we notice what happens to a permutation of n with no correct successors if we delete the item n. This will either leave an $(n-1)$ permutation with no correct successors or an $(n-1)$ permutation with exactly one correct successor (when the item n separated digits i, $i+1$ for some i). The latter case could arise from $u_{n-1,1}$ permutations and the former from $(n-1)u_{n-1,0}$ permutations, since the same $n-1$ permutation with no correct successors would arise from the deletion of the digit n from any of $n-1$ positions. Hence the relation is

$$u_{n,0} = (n-1)u_{n-1,0} + u_{n-1,1}$$

An application of the earlier relation for $u_{n,r}$ in the case $r = 1$ allows this two variable recurrence relation to be expressed as a single variable one in, say, u_n where $u_n = u_{n,0}$. The relation becomes

$$u_n = (n-1)u_{n-1} + (n-2)u_{n-2}$$

which is a second order difference equation with variable coefficients. If we propose to solve this equation using generating functions we might note that the problem from which it arose involves a number of permutations of n things and so we might expect the solution to increase rapidly, rather like $n!$; even the appearance of the equation itself which has coefficients linear in n occurring additively in the recurrence relation, also should suggest a quite rapid increase for u_n. It would therefore be prudent to define our generating function, using the set of linearly independent functions $\{x^n/n!\}$ instead of just the powers of x. Let $G(x) = \Sigma u_n x^n/n!$. The equation can be rewritten

$$n(n-1)\frac{u_n}{n!} = [(n-1)(n-2)+(n-1)]\frac{u_{n-1}}{(n-1)!} + (n-2)\frac{u_{n-2}}{(n-2)!}$$

which is now in a suitable form to be multiplied by x^n, and summed for $n = 0, 1, 2, \ldots$ The various terms are identified in terms of $G(x)$ and its derivatives as in the last section

$$x^2\frac{d^2G}{dx^2} = x^3\frac{d^2G}{dx^2} + x^2\frac{dG}{dx} + x^3\frac{dG}{dx}$$

This differential equation for the generating function $G(x)$ is apparently of second order but as $G(x)$ does not appear it can be treated as one of first order in $\frac{dG}{dx}$. It is thus of the 'variables separable' type, which integrates to

$$\log\frac{dG}{dx} = C - x - 2\log(1-x)$$

where C is an arbitrary constant. By exponentiating

$$\frac{dG}{dx} = \frac{Ae^{-x}}{(1-x)^2} = Ae^{-x}\{1 + 2x + 3x^2 + \ldots + (n+1)x^n + \ldots\}$$

where $A = e^C$.

There is of course no need to integrate again to find G from $\frac{dG}{dx}$ since $\frac{dG}{dx}$ itself is a generating function in the form

$$\frac{dG}{dx} = \sum_{n=1}^{\infty} \frac{u_n x^{n-1}}{(n-1)!}$$

and from its expansion we can pick out the appropriate coefficients u_n. If e^{-x} is expanded in powers of x and the coefficient of $x^{n-1}/(n-1)!$ picked out,

$$u_n = A(n-1)! \left[n - \frac{n-1}{1!} + \frac{n-2}{2!} - \ldots + \frac{(-1)^{n-1}}{(n-1)!} \right]$$

$$= A \left[n! \sum_{r=0}^{n-1} \frac{(-1)^r}{r!} + (n-1)! \sum_{r=0}^{n-2} \frac{(-1)^r}{r!} \right]$$

It only remains to identify the constant A ; since there is just one permutation of the first two integers which has no correct successors, namely (2 1), $u_2 = 1$ which gives $A = 1$. This completes the solution of the one variable equation by a generating function and gives the number of permutations with no correct successors in them $u_n = u_{n,0}$; this immediately allows the use of the relation between $u_{n,r}$ and $u_{n,0}$ to obtain the result for the number of permutations with a specified number of correct successors.

There is an interesting appearance of e , the base of natural logarithms, implicit in these results; $u_n/n!$ tends to e , and even for n quite small (for example, if n reaches double figures) the number of permutations of n with no correct successors is $\frac{n!}{e}(1 + \frac{1}{n})$ to a close approximation.

3.2 Reduction of the Order

One of the useful techniques available for solving linear differential equations applies the knowledge of one solution of the differential equation to derive another differential equation for the remaining solutions which will have one lower order than the original one. A similar technique is available for linear difference equations. Let us suppose that the equation to be solved is written in difference form as

$$a_k \Delta^k u_n + a_{k-1} \Delta^{k-1} u_n + \ldots + a_0 u_n = A_n$$

where the coefficients a_r and A_n may be arbitrary functions of the independent variable n. Suppose that g_n is known to be a solution to the homogeneous equation - that is, the insertion of g_n in place of u_n will cause the left hand side of the equation to vanish. The device used several times before lets $u_n = f_n g_n$ and seeks to derive an equation for the unknown function f_n.

It is convenient first to express the results of the finite difference operators Δ and E operating on u_n in terms of similar operators which are restricted to the functions f_n and g_n respectively. Let, for example, E_f be the shift operator restricted to f_n so that $E_f f_n = f_{n+1}$ while $E_f g_n = g_n$, with similar interpretations for Δ_f, E_g, Δ_g. Then

$$\Delta u_n = (E-1)u_n$$
$$= (E_f E_g - 1)f_n g_n$$
$$= \{(E_f - 1)E_g + E_g - 1\}f_n g_n$$
$$= (\Delta_f E_g + \Delta_g)f_n g_n$$

Thus there is a formal relation between these operators

$$\Delta \equiv \Delta_f E_g + \Delta_g \tag{3.6}$$

and this may be iterated to give

$$\Delta^k = (\Delta_f E_g + \Delta_g)^k \tag{3.7}$$

This relation can be used to substitute in the original equation for all the occurrences of Δ. Note that the last term of the binomial expansion for Δ^k is Δ_g^k. It follows that a collection of the terms in descending powers of Δ_f will give the coefficient of the zero power, i.e. of f_n itself, as

$$[a_k \Delta_g^k g_n + a_{k-1} \Delta_g^{k-1} g_n + \ldots + a_0 g_n]$$

This quantity within the square brackets is just the left hand side of the original equation (3.4) with the function g replacing the original unknown u ; and it has been supposed that g_n is known to be a solution to the homogeneous equation. Thus the coefficient of f_n after the expansion of the Δ operators will necessarily vanish, leaving an equation with terms only in $\Delta_f^k f_n, \Delta_f^{k-1} f_n, \ldots, \Delta_f f_n$. This is an equation of order just k-1 in the unknown variable $\Delta_f f_n$. The following example will illustrate the necessary operations in sequence. They are:

1. If the equation is not in Δ form, express it in this way.

2. Replace the unknown function by a product of a known solution to the homogeneous equation and a different unknown function.

3. Replace the general forward difference operator Δ by the appropriate equivalents in terms of the Δ's and E's restricted to the particular functions.

4. Obtain the equation of one order lower, checking that the coefficient of the new unknown function is indeed zero.

Example To solve

$$(2n-1)u_{n+2} - (8n-2)u_{n+1} + (6n+3)u_n = 0$$

The first task is to guess one solution of this equation. As its coefficients are all polynomials in n , the first search for a solution might be confined to polynomials. Thus we might assume $u_n = c_0 + c_1 n + \ldots$ In this case the constant c_0 alone fails to satisfy the equation but the next term enables us to do so and we find one solution $u_n = n$. With this knowledge the previous procedure can be applied to reduce the second order equation to a first order one and hence to an equation that can always be solved.

The first step is to express the equation in Δ form, using $E = 1 + \Delta$. The equation becomes

$$[(2n-1)(1+2\Delta+\Delta^2) - (8n-2)(1+\Delta) + 6n+3]u_n = 0$$

which simplifies to

$$[(2n-1)\Delta^2 - 4n\Delta + 4]u_n = 0$$

44

The next step replaces Δ by $\Delta_f E_\sigma + \Delta_\sigma$ and $u_n = f_n g_n$ where in this case $g_n = n$ so that $\Delta_g g_n = 1$, and $E_g g_n = n+1$, $E_g^2 g_n = n+2$, $\Delta_g^2 g_n = 0$. The equation therefore becomes

$$(2n-1)[(n+2)\Delta_f^2 + 2\Delta_f] f_n - 4n[(n+1)\Delta_f f_n + f_n] + 4nf_n = 0$$

which simplifies to

$$(2n-1)(n+2)\Delta_f^2 f_n - (4n^2+2)\Delta_f f_n = 0$$

There is some reassurance that an algebraic slip has been avoided by the vanishing of the coefficient of f_n and the derivation of a first order equation in the unknown $\Delta_f f_n = w_n$, say. There is no need now to retain the suffices on the Δ's and the solution of the first order equation in w_n can be derived. The reference to section 3.1.1 reminds us that for the solution of first order equations it is preferable to have the equation in the E or recurrence relation form rather than the Δ one. Substitution of the relation $\Delta = E-1$ yields

$$(2n-1)(n+2)w_{n+1} - (6n^2+3n)w_n = 0$$

Hence

$$w_{n+1} = \frac{3n(2n+1)w_n}{(n+2)(2n-1)}$$

Repeated evaluation of w_n in terms of w_{n-1}, w_{n-2}, etc. gives

$$w_{n+1} = \frac{3^n \cdot 2 \cdot (2n+1)}{(n+1)(n+2)} C$$

where C is constant. Since $w_n = f_{n+1} - f_n$ we can derive f_n by summation, using the convenient form of expression for

$$w_n = \frac{2C}{3}\left(\frac{3^{n+1}}{n+1} - \frac{3^n}{n}\right)$$

Thus

$$f_n = \frac{C_1 3^n}{n} + C_2$$

where C_1 and C_2 are arbitrary constants, C_2 appearing from the summation; which yields the solution of the original equation

$$u_n = f_n g_n = C_1 3^n + C_2 n$$

3.2.1 Application: Vacancies on a Line

A radio-chemist was performing an experiment directing hydrogen onto a surface. He had reason to believe that the molecules of hydrogen, each composed of two atoms, would form a film on the surface just one atom thick. He wished to consider a model of the process in which the surface was a rectangular lattice of sites which could be occupied by hydrogen atoms in such a way that a molecule would occupy two adjacent sites. In this model the molecules of hydrogen would come onto the surface one after another and the atoms would occupy sites and so make them unavailable to the atoms of molecules arriving later. The sequential occupation of the available sites by hydrogen atoms would isolate some single sites. These would have to remain unoccupied by hydrogen as they would have no adjacent partner to accommodate the molecule. The chemist wished to have some idea of the proportion of sites of the lattice which would have to remain vacant because of the chance distribution of the molecules. A two-dimensional model presented serious problems and, in an attempt to derive a useful approximation in the practical problem, attention was concentrated on a one-dimensional model, which took the following form:

1	2	$\cdots$	i	i+1	i+2				n-1	n

Consider a line of n possible sites which consists, of course, of n-1 adjacent pairs of sites. Let one of these n-1 pairs be selected at random with equal probability and occupied by a molecule. Further selection of

46

pairs proceeds sequentially, always by selecting at random one of the remaining pairs of adjacent sites for occupation until no further adjacent
pairs remain. Let u_n be the average number of isolated sites remaining
after this selection procedure has been completed when starting from a
line of n sites. The choice of the first pair for occupation divides the
line into two parts, say of length i and $n-i-2$ respectively (the first pair
selected occupying the sites $i+1$, $i+2$). After this selection the two parts
of the line are quite independent of one another and the average number of
isolated sites remaining in them will just be u_i and u_{n-i-2}. Therefore
we have the relation

$$u_n = \frac{1}{n-1} [(u_0 + u_{n-2}) + (u_1 + u_{n-3}) + \dots + (u_{n-2} + u_0)]$$

since the probability of any particular adjacent pair being selected first is
$1/(n-1)$. The relation can be written

$$(n-1)u_n = 2 \sum_0^{n-2} u_r$$

and a second order linear difference equation obtained by subtracting from
this the same relation with n replaced by $n-1$;

$$(n-1)u_n - (n-2)u_{n-1} = 2u_{n-2} \tag{3.8}$$

with the initial conditions $u_0 = 0$ and $u_1 = 1$. It is this second order
equation which we have to solve, and so look for one solution which can be
used to reduce its order. A trial solution of a first order polynomial once
again brings success; one solution of equation (3.8) is $u_n = a(n+2)$. The
technique of the previous section can now be applied and yields the solution
to the problem. The reader should check that

$$u_n = (n+2) \left[\frac{(-2)^{n+2}}{2(n+2)!} + \sum_{r=0}^{n+1} \frac{(-2)^r}{r!} \right]$$

It is interesting to note that this solution tends quite quickly to
$(n+2)/e^2$, once again introducing the base of natural logarithms in a
rather surprising way. The appearance of the factor $(n+2)$ has an
explanation in terms of the linear model. In many combinatorial problems

a simpler situation is obtained by ignoring 'end' or 'edge' effects and one
might try to eliminate these from the original model by imagining the line
of sites formed into a circle. If, however, there were $n+2$ sites in the
circle at the start, the selection of the first pair would immediately produce a straight line situation with n sites. Thus the circular problem
with $n+2$ sites is identical, as far as the average number of isolated
sites is concerned, with the linear problem of n sites and an intuitive
explanation of the factor $(n+2)$ appears.

3.3 PARTIAL DIFFERENCE EQUATIONS

Partial difference equations are those difference equations which
involve two or more independent variables; not surprisingly, general
equations of these types are not amenable to solution, but there are a few
special cases which can be solved and which are useful in certain applications.

If all the coefficients in a linear partial difference equation are
constants, the indicial equation approach for equations in one variable
can be generalised by assuming a trial solution $u_{m,n} = \alpha^m \beta^n$.

Example Solve $u_{m+1, n+1} = a u_{m,n}$

If the trial function $u_{m,n} = \alpha^m \beta^n$ is a solution,

$$\alpha\beta = a$$

Thus for any non-zero value of α, if $\beta = a/\alpha$, we have a solution to the
equation; since this constant coefficient equation is linear, any linear
combination of such solutions is still a solution. Thus the general solution
to our equation is

$$u_{m,n} = \sum A(\alpha, a/\alpha)\alpha^m a^n \alpha^{-n} = a^n \sum c_\alpha \alpha^{m-n}$$

where all the A's and c_α are arbitrary constants. In this form the
solution $u_{m,n}$ appears complicated but it is equivalent to

$$u_{m,n} = a^n f(m-n)$$

where f is an arbitrary function.

48

Another case on which an attack can be made is where the solution has the form of a product of two functions each of just one of the variables, i.e. when $u_{m,n} = f_m g_n$. The method involves collecting on each side of the equation terms in one of the variables only; each side of the equation then must be constant, and so give rise to two single variable difference equations which it might be possible to solve. The example of the next section displays the method in the context of a sorting problem.

3.3.1 A Sorting Application

Two strings of r, s items respectively are each in ascending order in the main store of a computer. The largest items of the strings are compared and the greatest placed in the last of a set of s locations which immediately follow those in which the r string is stored. The merging of the two strings continues from their largest ends until all of the s string has been moved; some of the r string may remain in their original places, correctly ordered in the final (r+s) string. Under the assumption that the members of the two strings are determined by a random selection of r from the whole r+s items, we find the average number of items that will not have to be moved in the store, during the complete merging.

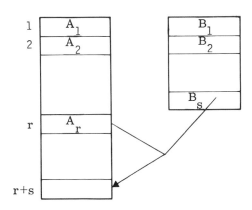

Fig. 3.1. Internal merging of two strings

The average number of items that will not have to be moved clearly depends upon the sizes r, s of the two strings. Let it be $u_{r,s}$. Since there are r items in the r string selected at random, out of the whole r+s items the probability that the largest item is in the r string is r/(r+s) . Once the largest item has been placed in position the problem is either that of merging an (r-1) string with an s string under the same conditions as the original problem, or of merging an r string with an (s-1) string. In those cases, the average number of items which would not have to be moved are respectively $u_{r-1,s}$ and $u_{r,s-1}$. Hence we must have

$$u_{r,s} = \frac{r}{r+s} u_{r-1,s} + \frac{s}{r+s} u_{r,s-1}$$

This is a two variable partial difference equation and has boundary conditions $u_{r,0} = r$ and $u_{0,s} = 0$, since if there is only an r string present all of its members will remain unmoved, while if there is just an s string every member will be transferred to a new place in the store. We seek a solution by assuming $u_{r,s} = f_r g_s$, hoping to find that the solution is of this form. The equation can be written

$$(r+s)f_r g_s = rf_{r-1}g_s + sf_r g_{s-1}$$

which can be rearranged as

$$r(f_r - f_{r-1})/f_r = -s(g_s - g_{s-1})/g_s$$

In this form the left hand side of the equation is a function of the variable r only, while the right hand side is a function of the variable s only. The only way in which such functions can be equal for all values of the variables concerned is that each is a constant, say k ; thus the original partial difference equation can be reduced to two linear difference equations each in one variable, namely

$$r(f_r - f_{r-1}) = kf_r$$

and

$$-kg_s = s(g_s - g_{s-1})$$

The boundary conditions $u_{r,0} = r$ and $u_{0,s} = 0$ allow us to take as boundary conditions for the single variable equations

$$f_r = r \, , \quad g_0 = 1$$

The solution of the equation for f is easily found to be

$$f_r = \frac{r! \, f_1}{(r-k)(r-1-k) \ldots (2-k)}$$

which reduces to $f_r = r$ when $k = 1$. With this value of k the equation for g yields

$$g_s = g_0/(s+1)$$

The solution of the partial difference equation is therefore

$$u_{r,s} = r/(s+1)$$

which is the average number of items in the r string which will not have to be moved in the merging process.

This result confirms the intuitive result that it is better to use the longer of the two strings as part of the reception area for the complete merged string; on average there would then be less items which have to be moved about the store, since $r/(s+1) > s/(r+1)$ if $r > s$.

An interesting alternative derivation of this result which is perhaps far easier to detect once one knows what the result is, is to observe that there are $s+1$ gaps in the ordered list of s items; an r-item selected at random from the same set as the s items will fall into the first gap with probability $1/(s+1)$, i.e. this is the probability of it being before all the s items. Accordingly $1/(s+1)$ is the probability that an item chosen at random does not need to be moved in the r-list; as there are r items, the average number which would not be moved is $r/(s+1)$.

3.4   **BIBLIOGRAPHY**

The general comments given in chapter 2 are also applicable to much of the material in this chapter. Milne-Thomson's sections 11.1, 11.3,

12.3, 13.8 are relevant. More extended treatment of the problem in section 3.2.1 is given in 'Distribution of Vacancies on a Line', E.S. Page, J. Roy. Statist. Soc. B, Vol. 21 (1959), pp364-74, and 'A Note on Vacancies on a Line', F. Downton, J. Roy. Statist. Soc. B, Vol. 23 (1961), pp207-14, while the application in section 3.3.1 is described in 'Another Note on Internal Merging', J.S. Clowes, E.S. Page and L.B. Wilson, Software Practice and Experience, Vol. 3 (1973), p185.

3.5 EXERCISES

[3.1] Solve $u_{n+1} = e^{2n} u_n$.

[3.2] Solve $u_{n+1} = (n+1)u_n - (n-1)!$ when $u_1 = 2$.

[3.3] Solve $(2n-1)u_{n+2} - (8n-2)u_{n+1} + (6n+3)u_n = 0$ using the known solution $u_n = 3^n$ to reduce the order by one (cf. section 3.2).

[3.4] Solve $(n+1)u_{n+2} - nu_{n+1} - u_n = 0$ by spotting one solution or by G.F.

[3.5] Solve $(n+2)u_{n+2} - nu_{n+1} - u_n = 0$.

[3.6] Solve $u_n = (n-1)(u_{n-1} + u_{n-2})$.

[3.7] In the model of vacancies on a line of section 3.2.1, derive and solve an equation for p_n , the probability that the left hand end point of the line remains vacant.

[3.8] Suppose the pairs have been placed on the line as in section 3.2.1, derive a difference equation for the number of different patterns of vacancies there can be in lines of n sites. What is the corresponding equation for the number of patterns when triplets instead of pairs are occupied sequentially?

[3.9] Solve $mu_{m-1,n} - nu_{m,n-1} = 0$, $m, n \geq 1$ and $u_{1,1} = 1$.

[3.10] A secretary types n letters and their corresponding envelopes; in a fit of aberration, she then puts the letters into the envelopes at random. What is the probability that no letter is in its correct envelope?

[3. 11] In the expansion of a determinant of order n, how many terms contain no diagonal element?

[3. 12] The number u_n of comparisons required to build a binary sequence search tree can be shown to satisfy the relation

$$u_n = u_{n-1} + (u_{n-1}+2n-1)/n$$

with $u_1 = 1$. Solve the equation for u_n, expressing u_n in terms of

$$H_n = \sum_{i=1}^{n} 1/i$$

What is the approximate rate of growth of u_n for large n?

(Newcastle 1977)

4 · Elementary Configurations

INTRODUCTION

Combinations, permutations, partitions, compositions, and graphs are the simplest and the most widely used mathematical objects in combinatorics. Since the mathematical properties of these configurations are so well known we will give in this chapter only the basic definitions and techniques and the reader should consult other books, for example those by Liu and Riordan, for further details. This chapter thus outlines the background for the computational problems associated with the efficient generation of these elementary configurations by computer, which will be discussed in chapter 5 when the wider concepts of ordering are introduced.

4.2 COMBINATIONS AND PERMUTATIONS

Definitions A combination of n objects taken r at a time (called an r-combination of n elements) is a selection of r of the objects where the order of the objects in the selection is immaterial.

A permutation of n objects taken r at a time (called an r-permutation of n elements) is an ordered selection of r of the objects.

These definitions say nothing about whether the objects may appear more than once; unlimited repetition of objects can be allowed or all repetition forbidden or there can be any rule between these extremes. Let us now look at some simple examples of combinations and permutations.

Suppose there are four objects a, b, c and d and selections of them are made two at a time. The combinations without repetition are six in number, namely

    a b, a c, a d, b c, b d, c d

whilst there are ten with repetition, which are

$$aa, \ ab, \ ac, \ ad, \ bb, \ bc, \ bd, \ cc, \ cd, \ dd$$

The twelve permutations without repetition are

$$ab, \ ac, \ ad, \ ba, \ bc, \ bd, \ ca, \ cb, \ cd, \ da, \ db, \ dc$$

whilst the sixteen with repetition are

$$aa, \ ab, \ ac, \ ad, \ ba, \ bb, \ bc, \ bd, \ ca, \ cb, \ cc, \ cd, \ da, \ db, \ dc, \ dd$$

4.2.1 The Enumerators for Permutations and Combinations

The general formulae for enumerating permutations and combinations will now be considered. If we start with permutations without repetition and examine the r-permutation $x_1 x_2 \ldots x_r$ obtained from the n objects $\{x_i\}$ $(i=1, 2, \ldots, n)$ then we may choose one of the $\{x_i\}$ to be x_1 in n ways; having done so we may choose x_2 from the $\{x_i\}$ which remain in n-1 ways, each of which may be combined with every one of the n possible x_1 and so on; in general the kth element in the permutation, x_k, may be chosen in only n-k+1 ways. Therefore the number of r-permutations of n objects, $P(n, r)$, is given by

$$P(n, r) = n(n-1) \ldots (n-r+1) \tag{4.1}$$

$$= \frac{n!}{(n-r)!} \tag{4.2}$$

One approach to this result via difference equations notes that once the first object, x_1, has been selected, any (r-1)-permutation of the remaining n-1 $\{x\}$ objects may be attached to x_1 to give the r-permutation of n objects. Hence

$$P(n, r) = nP(n-1, r-1)$$

or, in more familiar suffix notation,

$$p_{n,r} = np_{n-1,r-1} \qquad (4.3)$$

When explicit reference to r is not made we assume that all the objects are to be permuted; thus when we talk about 'the permutations of n objects' we mean the case $r = n$. In this case the solution of the one variable difference equation $p_{n,n} = np_{n-1,n-1}$ is $p_{n,n} = P(n,n) = n!$.

When unlimited repetition in an r-permutation of n objects is allowed then after choosing the first object in n ways, the next object can also be chosen in n ways and so on. Therefore in this case

$$U(n,r) = n \times n \times \ldots \times n$$
$$= n^r \qquad (4.4)$$

There are, of course, intermediate cases between no repetition and unlimited repetition of the objects. Consider for example n objects of which m_1 are of the first kind, m_2 are of the second kind, ..., m_k are of the kth kind, so that

$$\sum_{i=1}^{k} m_i = n$$

The number of permutations of all the objects in this case is

$$Y = \frac{n!}{m_1! m_2! \ldots m_k!} \qquad (4.5)$$

This result can be derived by first considering all the like objects replaced by new distinct objects, which gives n! permutations. The m_1 like objects of the first kind have $m_1!$ permutations of themselves in their positions in the whole n-permutation which result in $m_1!$ indistinguishable n-permutations. Similarly for the other sets of like objects. Therefore $n! = Y m_1! m_2! \ldots m_k!$ which yields (4.5).

If we now turn to combinations and consider the number of r-combinations of n objects without repetition, we note that the r objects of each r-combination can be permuted among r! different r-permutations each of which, of course, corresponds to a single combination. Hence if the number of r-combinations of n objects without repetition is denoted by $C(n,r)$,

$$r!\,C(n, r) = P(n, r) = \frac{n!}{(n-r)!}$$

$$C(n, r) = \frac{n!}{r!\,(n-r)!} = \binom{n}{r} \tag{4.6}$$

The last symbol is that usually used for the binomial coefficient, i.e. it is the coefficient of x^r in the expansion of $(1+x)^n$; it is, of course, the number of ways of selecting r out of the n brackets $(1+x)(1+x)\ldots(1+x)$ to use the 'x' term in the multiplication.

There are many different difference equations for the number of r-combinations of n objects; one of the fundamental ones can be justified by focussing attention on a particular one of the n objects, and considering how many of the r-combinations contain it (namely $C(n-1, r-1)$ since the other $r-1$ objects must be selected from the remaining $n-1$) and how many do not (in this case $C(n-1, r)$ since the r objects can be selected from only $n-1$). Hence

$$C(n, r) = C(n-1, r-1) + C(n-1, r) \tag{4.7}$$

which is the familiar Pascal triangle relation.

A possible representation of a combination is as a binary number with n bits. If we have eight objects $(1, 2, 3, 4, 5, 6, 7, 8)$ taken three at a time then the 3-combination $(2\ 6\ 7)$ could be represented by the binary number

01100010

where there are ones in the second, sixth and seventh positions from the least significant end. If we read this number in the same manner but from left to right, we have the combination $(2\ 3\ 7)$, sometimes called the reverse combination, whilst if we interchange zeros and ones we have the combination $(1\ 3\ 4\ 5\ 8)$, sometimes called the dual combination. Clearly there are the same numbers of them which is confirmed by the formula since $C(n, r) = C(n, n-r)$.

The number of binary numbers with n bits and r of those bits equal to one is, of course, $\binom{n}{r}$.

The general formula for r-combinations of n objects when unlimited repetition is allowed is more difficult to obtain - we cannot simply divide the permutation result for unlimited repetition, n^r , by an appropriate factor since different combinations with repetition will not in general give rise to the same number of permutations. For example (a a b) gives rise to three different permutations while (a b c) gives six permutations.

The classical method of arriving at the correct formula is as follows:

Let one r-combination of n objects (which are considered to be the digits $1, 2, 3, \ldots, n$) in which repetition is allowed be $(c_1\ c_2 \ldots c_r)$, and suppose $c_1, c_2, \ldots, c_r$ are in rising order, i.e. $c_1 \leq c_2 \leq \ldots \leq c_r$. Form the set of d's , $d_1, d_2, \ldots, d_r$, by the rule $d_1 = c_1 + 0$, $d_2 = c_2 + 1$, $\ldots, d_i = c_i + i - 1, \ldots, d_r = c_r + r - 1$. This transformation ensures that the d's are unalike whatever the c's were. It is clear that the sets of c's and d's are equinumerous since every distinct r-combination of the c's produces a distinct set of d's and vice versa. The number of sets of d's is the number of r-combinations without repetition of the objects $1, 2, \ldots, n+r-1$ since the largest d is d_r when c_r has its maximum value n . Thus the number of sets of d's is $\binom{n+r-1}{r}$ and this is equal to the number of r-combinations of n objects with unlimited repetition.

If $V(n, r)$ is the number of unrestricted r-combinations of n objects, the number of them that contain at least one occurrence of a particular object is $V(n, r-1)$ (since one of those selected has been determined while the remaining r-1 are to be selected without restriction from the same n objects). The number out of the $V(n, r)$ that do not contain the particular object at all is $V(n-1, r)$ since only (n-1) objects are available for selection. Hence

$$V(n, r) = V(n, r-1) + V(n-1, r) \tag{4.8}$$

the appropriate solution of which is

$$V(n, r) = \binom{n+r-1}{r} \tag{4.9}$$

as before.

A bit pattern representation for r-combinations with repetition is not quite so simple as before; it needs to contain n+r-1 bits, again usually read from right to left. Taking our previous example of a 3-combination of the eight digits $(1, 2, \ldots, 8)$ then $(3\ 5\ 5)$ is represented by the binary number

$$0001100100 \qquad\qquad (4.10)$$

The least significant zero represents the fact that there are no ones in the combination, the next zero that there are no twos, the next one then represents the three and the zero after it the fact that there is only one three, and so on.

The construction of adding $0, 1, 2, 3, \ldots$ to a combination with repetition (which was used in the basic theorem above) can also be used in the bit pattern representation. Thus in our example $(3\ 5\ 5)$ becomes $(3\ 6\ 7)$ which we can see is the binary number (4.10) interpreted as a combination without repetition of ten bits.

4.2.2 Distributions

A distribution is defined as a separation of a set of objects into a number of classes - for example, the assignment of the objects to cells (or boxes); problems about distributions are very closely related to problems of permutations and combinations.

Consider first the case of assigning r different objects to n distinct cells such that each cell has at most one object. If $n \geq r$ then there are $P(n, r)$ ways, since the first object may be assigned to any of the n cells, the second object to one of the $n-1$ remaining cells etc. Alternatively if $r \geq n$ then there are $P(r, n)$ ways, since the object assigned to the first cell may be done in r ways, the object assigned to the second cell in $r-1$ ways etc.

Continuing with r different objects and n distinct cells but now allowing each cell to hold any number of objects, we obtain n^r ways of distributing the objects. This is true whether n is larger or smaller than r since the first object can be assigned to any one of the n cells and so can the second and the other objects.

When the r objects to be distributed are not all different suppose that m_1 of them are of the first kind, m_2 of the second kind, $\ldots$, m_k of them of the kth kind, so that $r = \sum_{i=1}^{k} m_i$. First suppose that each of the n distinct cells may hold at most one object $(n \geq r)$. The r cells are selected from the n cells (in $C(n, r)$ ways) and then the r objects are distributed into these r cells which is equivalent to forming a permutation with repetition of the objects. Using equation (4.5) there are $\frac{r!}{m_1! m_2! \ldots m_k!}$ such permutations. Therefore the number of these distributions is

$$D_1(n, r) = C(n, r) \frac{r!}{m_1! m_2! \ldots m_k!} = \frac{n!}{(n-r)! m_1! m_2! \ldots m_k!} \qquad (4.11)$$

When the r objects are all alike then there is only one kind, and $m_1 = r$. Thus the result given in (4.11) becomes $\frac{n!}{(n-r)! r!} = C(n, r)$.

The more interesting case is when the r like objects are placed in n distinct cells without any restriction on the number going into each cell. The number of ways of doing this is equivalent to selecting r cells from n with repetition of cells allowed and from equation (4.9) the number of such distributions is $C(n+r-1, r)$.

Many other variations are possible, some of which are given as exercises at the end of this chapter, and others are described in the books mentioned in the bibliography.

4.3 GENERATING FUNCTIONS

In a previous section (2.4) generating functions were introduced as a means of describing infinite sequences of numbers; in that chapter they were used to solve difference equations. Here they are applied to the enumeration of permutations and combinations.

Consider the three distinct objects a, b and c, and form the polynomial

$$(1+ax)(1+bx)(1+cx) = 1 + (a+b+c)x + (ab+bc+ca)x^2 + abcx^3 \qquad (4.12)$$

The coefficients of x on the right hand side have some interesting properties; if we consider the three ways of selecting one object (a or b or c) and represent it a+b+c then we have the coefficient of the first power of x, $x^1 = x$. Similarly the three ways of selecting two objects (ab or bc or ca) may be represented ab+bc+ca which is the coefficient of x^2. Finally the single way of selecting all three objects, namely abc, is the coefficient of x^3. This result is not accidental; on the left hand side of equation (4.12) the factor $1+ax$ can be considered as representing symbolically the two ways of selecting a or not, the 1 or x^0 representing the 'non-selection of a' and the ax representing the 'selection of a'. The factors $(1+bx)$ and $(1+cx)$ can be interpreted in a similar manner. Thus the product of the three factors $(1+ax)(1+bx)(1+cx)$ indicates the selection or non-selection of all three objects a, b and c, and the powers of x in the product indicate the number of objects selected. Thus the coefficient of x^2 is an enumeration of the ways in which we can select two objects. The coefficients in this example can be interpreted literally (i.e. by the letters) as listing the various selections, and numerically, by placing $a=b=c=1$, as the numbers of those selections.

4.3.1 Combinations

The polynomial $(1+ax)(1+bx)(1+cx)$ given in equation (4.12) was shown to represent the different ways of selecting the three objects a, b and c. This can be generalised to n objects, say $a_1, a_2, \ldots, a_n$, by expanding in powers of x the polynomial

$$
\begin{aligned}
&(1+a_1 x)(1+a_2 x)(1+a_3 x) \ldots (1+a_n x) \\
&= 1 + (a_1 + a_2 + \ldots + a_n)x + (a_1 a_2 + a_1 a_3 + \ldots)x^2 + \ldots
\end{aligned}
\tag{4.13}
$$

The coefficient of x^r on the right hand side represents the r-combination chosen from the objects $a_1, a_2, \ldots, a_n$. However, we often require the number of r-combinations of n objects rather than the actual combinations. In this case we can use the above formula with $a_1 = a_2 = a_3 = \ldots = a_n = 1$. In the simple case of three objects we obtain

$$(1+x)^3 = 1 + 3x + 3x^2 + x^3$$

whilst for n objects

$$(1+x)^n = 1 + nx + \frac{n(n-1)}{2}x^2 + \ldots + \frac{n(n-1)(n-2)\ldots(n-r+1)x^r}{r!} + \ldots + x^n$$

$$= C(n,0) + C(n,1)x^1 + C(n,2)x^2 + \ldots + C(n,r)x^r + \ldots + C(n,n)x^n$$

$$\tag{4.14}$$

which is the familiar binomial expansion.

A generating function used in this way is usually called an enumerator. To illustrate the use of enumerators consider the following example.

How many combinations of three objects can be formed if one object can be selected at most once, the second object at most twice, and the third object at most three times?

Since the first object can appear either once or not at all, the factor in the enumerator must contain x^0 and x^1 but no other terms; similarly the factor for the second object will have x^2 appearing as well. The enumerator can thus be written down one factor at a time corresponding to each of the objects.

The enumerator is

$$(1+x)(1+x+x^2)(1+x+x^2+x^3) = 1 + 3x + 5x^2 + 6x^3 + 5x^4 + 3x^5 + x^6$$

If an object is allowed unlimited repetition the corresponding factor in the enumerator must have every power of x present and so is

$$(1 + x + x^2 + \ldots + x^i + \ldots) = (1-x)^{-1}$$

Thus the enumerator of r-combinations of n objects with unlimited repetition is

$$(1-x)^{-n} = 1 + nx + \frac{n(n+1)}{2!} x^2 + \ldots + \frac{n(n+1)\ldots(n+r-1)}{r!} x^r + \ldots$$

$$= \sum_{r=0}^{\infty} \binom{n+r-1}{r} x^r \qquad (4.15)$$

This result was derived previously (equation (4.9)) by traditional methods.

These examples have shown that generating functions are a powerful tool for enumerating combinations with various kinds of repetitions and it is natural to turn to permutations and see if they can help there.

4.3.2 Permutations

In ordinary algebraic multiplication it is not possible to distinguish between ab and ba , because multiplication is commutative, ab=ba . This is a snag when one tries to develop a generating function for permutations. In the case of three objects a, b and c , the generating function desired has the form

$$1+(a+b+c)x + (ab+ba+bc+cb+ca+ac)x^2 + (abc+acb+bac+bca+cab+cba)x^3$$

$$= 1+(a+b+c)x + 2(ab+bc+ca)x^2 + 6abcx^3$$

In the third term on the right the distinction has been lost between the permutations ab and ba . It would, of course, be possible to develop a non-commutative algebra but simplicity would be lost. We therefore try an alternative strategy that stays in the field of real numbers and ordinary algebra.

Consider the binomial expansion

$$(1+x)^n = 1+C(n,1)x + C(n,2)x^2 + \ldots + C(n,r)x^r + \ldots + x^n$$

and replace $C(n,r)$ by $P(n,r)/r!$

$$(1+x)^n = 1 + \frac{P(n,1)}{1!}x + \frac{P(n,2)}{2!}x^2 + \ldots$$

$$+ \frac{P(n,r)}{r!}x^r + \ldots + \frac{P(n,n)}{n!}x^n \qquad (4.16)$$

In this expansion $P(n,r)$ is the coefficient of $x^r/r!$ and it appears that the appropriate functions in the expansion for permutations are the powers of x reduced by the factorial. We thus define another kind of generating function, the <u>exponential generating function</u>,

$$E(x) = a_0 + a_1\frac{x}{1!} + a_2\frac{x^2}{2!} + \ldots + a_i\frac{x^i}{i!} + \ldots \qquad (4.17)$$

$$= \sum_{i=0}^{\infty} a_i\frac{x^i}{i!}$$

which again is a convenient way of summarising the infinite sequence $\{a_i\}$. It has appeared previously in a different guise when solving certain difference equations whose solutions increased rapidly (section 3.1.3).

When the result given in equation (4.16) is generalised to enumerate r-permutations of n objects with unlimited repetition then the factor for each object must represent the fact that the object may not appear, or appear once, twice, etc. Thus each factor is

$$(1 + \frac{x}{1!} + \frac{x^2}{2!} + \frac{x^3}{3!} + \ldots + \frac{x^i}{i!} + \ldots)$$

and the exponential enumerator is

$$(1 + \frac{x}{1!} + \frac{x^2}{2!} + \frac{x^3}{3!} + \ldots)^n = e^{nx}$$

$$= \sum_{r=0}^{\infty} n^r \frac{x^r}{r!} \qquad\qquad (4.18)$$

a result which was derived previously in equation (4.4).

4.4   THE PRINCIPLE OF INCLUSION AND EXCLUSION

The principle of inclusion and exclusion (sometimes known as the sieve method) is an important combinatorial counting theorem. Consider first a simple illustrative example: A golf club has 125 members, 89 are male and 18 are left-handed players, 11 of the men also play left-handed. How many right-handed lady golfers does the club have?

The answer can be written $125-89-18+11 = 29$. We start with all the members and subtract from it those who are male and those who are left-handed. At this stage we have subtracted twice the left-handed males and so we must add them back in again to get the correct final answer. Diagrammatically it looks as follows:

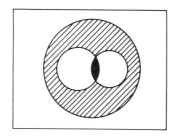

The whole area inside the large circle represents the members of the golf club. The areas inside the two smaller circles represent respectively the males and left-handed players so that the black area represents the left-handed males; this is the area which is subtracted twice and which must therefore be added back.

Generalising this example, consider N objects and the two properties a and b ; then the number without both these properties is given by

$$N(a',b') = N - N(a) - N(b) + N(ab)$$

for in subtracting from N those with property a and those with property b we have subtracted those with both properties $(N(ab))$ twice and they must be replaced. The general result for r properties is:

Theorem

If, of N objects, $N(a_1)$ have the property a_1 , $N(a_2)$ the property $a_2, \ldots,$ $N(a_r)$ the property a_r , $N(a_1 a_2)$ both the properties a_1 and $a_2, \ldots,$ $N(a_1 a_2 \ldots a_r)$ all the properties $a_1, a_2, \ldots, a_r$, then the number $N(a_1' a_2' \ldots a_r')$ with none of these properties is given by

$$
\begin{aligned}
N(a_1' a_2' \ldots a_r') = {} & N - N(a_1) - N(a_2) - \ldots - N(a_r) \\
& + N(a_1 a_2) + N(a_1 a_3) + \ldots + N(a_{r-1} a_r) \\
& - N(a_1 a_2 a_3) - \ldots - N(a_{r-2} a_{r-1} a_r) \\
& + \ldots + (-1)^r N(a_1 a_2 \ldots a_r)
\end{aligned}
\tag{4.19}
$$

Proof The theorem will be proved by induction on the number of properties. It is obviously true for one property since

$$N(a_1') = N - N(a_1)$$

Assume it is true for the $(i-1)$ properties $a_1, a_2, \ldots, a_{i-1}$ so that

$$
\begin{aligned}
N(a_1' a_2' \ldots a_{i-1}') = {} & N - N(a_1) - N(a_2) \ldots - N(a_{i-1}) \\
& + N(a_1 a_2) + \ldots + N(a_{i-2} a_{i-1}) \\
& + \ldots + (-1)^{i-1} N(a_1 a_2 \ldots a_{i-1})
\end{aligned}
\tag{4.20}
$$

In the above equation replace the collection N by the collection $N(a_i)$, then

$$N(a_1' a_2' \ldots a_{i-1}' a_i) = N(a_i) - N(a_i a_1) - N(a_i a_2) - \ldots - N(a_i a_{i-1})$$
$$+ N(a_1 a_2 a_i) + \ldots$$
$$+ (-1)^{i-1} N(a_1 a_2 \ldots a_{i-1} a_i) \qquad (4.21)$$

Now the formula

$$N(x') = N - N(x) \qquad (4.22)$$

can be applied to any collection of objects suitably defined. Let N be the collection of objects without the properties $a_1, a_2, \ldots, a_{i-1}$ and x be the property a_i, then (4.22) becomes

$$N(a_1' a_2' \ldots a_{i-1}' a_i') = N(a_1' a_2' \ldots a_{i-1}') - N(a_1' a_2' \ldots a_{i-1}' a_i)$$

The two quantities on the right hand side are given by equations (4.20) and (4.21) and when these equations are subtracted we get the result given in the statement of the theorem with i for r. So the induction proof is complete.

Example 1 How many of the first 1000 integers are not divisible by 2, 3, 5 or 7?

Let a_1, a_2, a_3 and a_4 be the properties of divisibility by 2, 3, 5 and 7 respectively.

$N(a_1) = 500, \ N(a_2) = 333, \ N(a_3) = 200, \ N(a_4) = 142$
$N(a_1 a_2) = 166, \ N(a_1 a_3) = 100, \ N(a_1 a_4) = 71, \ N(a_2 a_3) = 66$
$N(a_2 a_4) = 47, \ N(a_3 a_4) = 28$
$N(a_1 a_2 a_3) = 33, \ N(a_1 a_2 a_4) = 23, \ N(a_1 a_3 a_4) = 14, \ N(a_2 a_3 a_4) = 9$
$N(a_1 a_2 a_3 a_4) = 4$
$N(a_1' a_2' a_3' a_4') = 1000 - 500 - 333 - 200 - 142 + 166 + 100 + 71 + 66 + 47 + 28$
$$- 33 - 23 - 14 - 9 + 4 = 228$$

Example 2 Derangements (problème des rencontres).

How many permutations of the n distinct elements $(1, 2, 3, \ldots, n)$ are there in which the element k is not in the kth position?

Let $b_1 b_2 \ldots b_n$ be a permutation of $1, 2, \ldots, n$. Let the property that $b_i = i$ be a_i.

Then for any s properties $a_{i_1}, a_{i_2}, \ldots, a_{i_s}$

$$N(a_{i_1} a_{i_2} \ldots a_{i_s}) = (n-s)!$$

because if s positions are fixed the remaining positions give $(n-s)!$ permutations.

Then the principle of inclusion and exclusion gives

$$N(a'_1 a'_2 \ldots a'_n) = n! - n(n-1)! + \binom{n}{2}(n-2)! + \ldots$$
$$+ (-1)^i \binom{n}{i}(n-i)! + \ldots + (-1)^n$$
$$= n! \, (1 - \frac{1}{1!} + \frac{1}{2!} - \ldots + \frac{(-1)^i}{i!} + \ldots + \frac{(-1)^n}{n!})$$
$$\simeq \frac{n!}{e} \quad \text{if} \ n \ \text{large}.$$

For $n = 4$ the derangements number $9 = 12-4+1$, and they are the permutations 2143, 2341, 2413, 3142, 3412, 3421, 4123, 4312, 4321.

4.5 PARTITIONS AND COMPOSITIONS OF INTEGERS

Partitions are first supposed to have appeared in a letter from Leibniz to Bernoulli in 1669, but as with so much else in combinatorics, it was Euler who did most of the early development. Some of the main results are given in this section but those who wish to find out more should consult MacMahon's book.

It is clearly possible to represent any positive integer n as a sum of one or more positive integers (a_i)

$$n = a_1 + a_2 + \ldots + a_m \tag{4.23}$$

Interesting problems arise in the enumeration and efficient generation of such representations. Divisions of a positive integer as in (4.23) are of two types depending on whether the ordering of the parts $a_1, a_2, \ldots, a_m$ is regarded as important or not. Ordered divisions are called compositions while unordered divisions are called partitions. This distinction is similar to that made between permutations and combinations. Compositions have attracted much less attention from mathematicians than partitions mainly because the theory about them appears straightforward and thus less challenging. However they seem to occur in practical situations as often as do partitions. Another distinction that can be made in divisions such as (4.23) is whether the number of parts is stated as m, say, or is un-specified.

Consider, for example, the partitions and compositions of the integer 5. There are seven unrestricted partitions, namely

5, 4+1, 3+2, 3+1+1, 2+2+1, 2+1+1+1, 1+1+1+1+1

and two of these, namely 4+1 and 3+2, have exactly two parts.

There are sixteen unrestricted compositions of $n = 5$

5, 4+1, 1+4, 3+2, 2+3, 3+1+1, 1+3+1, 1+1+3, 2+2+1, 2+1+2,
1+2+2, 2+1+1+1, 1+2+1+1, 1+1+2+1, 1+1+1+2, 1+1+1+1+1

and four of these have exactly two parts.

In the rest of this book when we write partitions or compositions we will omit the $+$ signs, thus 2+1+1+1 will be written 2111, or 21^3, and for partitions the largest parts will be written first.

4.5.1 Compositions

One of the easiest ways of enumerating the unrestricted compositions of n is to consider n ones in a row. Since there is no restriction on the number of parts, we may or may not put a marker in any of the $(n-1)$ spaces between the ones in order to form groups; this may be done in 2^{n-1} ways. The same type of argument can be applied when we restrict the compositions to have exactly m parts. Just $(m-1)$ markers are needed to form m groups and the number of ways of placing $(m-1)$ markers in the $(n-1)$ spaces between the ones is $\binom{n-1}{m-1}$.

These results can also be obtained as follows using generating functions. Let $C_m(x)$ be the enumerator for compositions of n with exactly m parts, where

$$C_m(x) = \sum_n C_{mn} x^n \tag{4.24}$$

and C_{mn} , the coefficient of x^n in this series, is the number of compositions of n into exactly m parts. Each part of any composition can be one, two, three or any greater number so that the factor in the enumerator must contain each of these powers of x , and so is

$$x + x^2 + x^3 + x^4 + \ldots + x^k + \ldots = x(1-x)^{-1}$$

Since there are exactly m parts, the generating function is the product of m such factors:

$$C_m(x) = (x+x^2+x^3+\ldots+x^k+\ldots)^m \tag{4.25}$$

which can be rewritten

$$C_m(x) = x^m(1-x)^{-m} = x^m \sum_{i=0}^{\infty} \binom{m+i-1}{i} x^i$$

Replacing $m+i$ by r in the summation

$$C_m(x) = \sum_{r=m}^{\infty} \binom{r-1}{r-m} x^r = \sum_{r=m}^{\infty} \binom{r-1}{m-1} x^r$$

so that the coefficient of x^n in this enumerator is $\binom{n-1}{m-1}$, as before.

The enumerating generating function for compositions with no restriction on the number of parts $C(x)$, can be obtained from $C_m(x)$ by summing

$$C(x) = \sum_{m=1}^{\infty} C_m(x) = \sum_{m=1}^{\infty} x^m(1-x)^{-m}$$

Substituting $t = x/(1-x)$ in the series we obtain

$$C(x) = (t+t^2+t^3+\ldots)$$

$$C(x) = \frac{t}{1-t} = \frac{x}{1-2x} = \sum_{n=1}^{\infty} 2^{n-1} x^n \tag{4.26}$$

Since the coefficient of x^n in the enumerator is 2^{n-1} this yields again the number of unrestricted compositions of n.

Many other results for compositions can be obtained by suitably modifying the two basic generating functions $C_m(x)$ and $C(x)$ given in equations (4.25) and (4.26).

4.5.2 Partitions

We noticed in deriving equation (4.6) that the number of r-combinations could be obtained from the number of r-permutations. Unfortunately no such simple relationship exists between the number of partitions and the number of compositions, because each partition will in general give rise to a different number of compositions. For example the two partitions of 10, 811 and 4321, give respectively three and twenty-four compositions. It is therefore impossible to use the results obtained in the previous section for compositions and we need to start again from basic concepts to derive a generating function $p(x)$ for partitions.

Let p_n be the number of unrestricted partitions of n so that the generating function is

$$p(x) = p_0 + p_1 x + p_2 x^2 + \ldots + p_n x^n + \ldots \tag{4.27}$$

Consider the polynomial

$$1 + x + x^2 + x^3 + \ldots + x^k + \ldots + x^n$$

The appearance of x^k can be interpreted as the existence of just k ones in a partition of the integer n. Similarly the polynomial

$$1 + x^2 + x^4 + \ldots + x^{2k} + \ldots$$

is concerned with the twos in the partition, and, in particular, the coefficient of $x^{2k} = (x^2)^k$ represents the case of just k twos in the partition. In general the polynomial

$$1 + x^r + x^{2r} + \ldots + x^{kr} + \ldots$$

can represent the r's in the partition. The generating function will need one factor for the ones, another factor for the twos, and so on. Collecting together these polynomials, the generating function for the partitions of n is obtained as

$$p(x) = (1+x+x^2+x^3+\ldots+x^k+\ldots)(1+x^2+x^4+\ldots+x^{2k}+\ldots)$$

$$\ldots (1+x^r+x^{2r}+\ldots+x^{kr}+\ldots) \ldots \tag{4.28}$$

$$= (1-x)^{-1}(1-x^2)^{-1} \ldots (1-x^r)^{-1} \ldots \tag{4.29}$$

The number of unrestricted partitions of n is therefore the coefficient of the term x^n in equation (4.29). Unfortunately we cannot obtain a simple formula for these coefficients but this does not mean that the generating function approach to partitions serves no useful purpose. Many of the standard results in partitions can be proved using them; for example, proving that the partitions with unequal parts are equinumerous with the partitions in which all parts odd is very simply done this way (see exercise (4.17)). The generating functions for many of the restricted partitions can also be derived from the basic forms given in equations (4.28) and (4.29); the factors in the generating function will be amended by the types of restriction imposed.

Examples
(a) The partitions with every part odd are enumerated by the generating function containing only the factors corresponding to ones, threes, fives etc. ; thus by a derivation similar to (4.29)

$$0(x) = (1-x)^{-1}(1-x^3)^{-1}(1-x^5)^{-1} \ldots \tag{4.30}$$

(b) The enumerator for the partitions with unequal parts is obtained by remembering that each factor can only contain two terms, one which indicates that the corresponding number does not appear in the partition and the other that the number appears once

$$u(x) = (1+x)(1+x^2)(1+x^3) \ldots (1+x^k) \ldots \tag{4.31}$$

71

(c) The enumerator for partitions in which no part is greater than s is

$$p_s(x) = (1-x)^{-1}(1-x^2)^{-1} \ldots (1-x^s)^{-1} \qquad (4.32)$$

since the factors commencing with higher powers than x^s are excluded.

When the partitions are to be enumerated by the number of parts, the basic generating function is more complex. We can start with a two variable generating function which was first used by Euler

$$F(x, v) = (1+vx+v^2x^2+\ldots)(1+vx^2+v^2x^4+\ldots)$$

$$\ldots (1+vx^i+v^2x^{2i}+\ldots) \ldots$$

The first factor in this G. F. represents $0, 1, 2, \ldots$ appearances of the part 1. The general term $v^j x^{ji}$ in the ith factor is that which indicates j appearances of the part i and it is the power of v which shows the number of parts. This two variable G. F. can also be written

$$F(x, v) = (1-vx)^{-1}(1-vx^2)^{-1} \ldots (1-vx^i)^{-1} \ldots \qquad (4.33)$$

If $p(x, k)$ is the enumerator for partitions with exactly k parts then

$$F(x, v) = \sum p(x, k)v^k \qquad (4.34)$$

and we can derive $p(x, k)$ as follows. From equation (4.33)

$$(1-vx)F(x, v) = (1-vx^2)^{-1}(1-vx^3)^{-1} \ldots (1-vx^i)^{-1} \ldots$$

$$= F(x, vx)$$

From equation (4.34)

$$(1-vx) \sum p(x, k)v^k = \sum p(x, k)v^k x^k$$

Hence,

72

$$p(x, k) - xp(x, k-1) = x^k p(x, k)$$

by equating coefficients of v^k . This is a first order difference equation in the index k with initial value $p(x, 0) = 1$, and is solved as in section 3.1.1 to give

$$p(x, k) = x^k (1-x)^{-1}(1-x^2)^{-1} \ldots (1-x^k)^{-1} \tag{4.35}$$

The coefficient of x^n in this generating function is the number of partitions of n with exactly k parts.

An alternative derivation of equation (4.35) is to use the enumerator for partitions with no part greater than k (since this is equal to the enumerator for partitions with at most k parts) and subtract from it the enumerator for partitions with no part greater than $k-1$. Using equation (4.32) this gives

$$\begin{aligned} p(x, k) &= \frac{1}{(1-x)(1-x^2) \ldots (1-x^k)} - \frac{1}{(1-x)(1-x^2) \ldots (1-x^{k-1})} \\ &= x^k (1-x)^{-1}(1-x^2)^{-1} \ldots (1-x^k)^{-1} \end{aligned}$$

as before.

4.5.3 The Ferrers Graph

Partitions may often conveniently be represented by an array of dots known as the Ferrers graph. For example, the Ferrers graph for the partition 5 3 2 2 (which is in standard form with the largest part first) is

A Ferrers graph has the following properties:

(i)   There is one row for each part.

(ii)   The number of dots in a row is the size of that part.

(iii)   An upper row always contains at least as many dots as a lower row.

(iv)   The rows are aligned on the left.

The partition obtained by reading the Ferrers graph by columns is called the conjugate partition. In the example above the conjugate partition of 5 3 2 2 is 4 4 2 1 1 . A partition whose Ferrers graph reads the same by rows and by columns is called self-conjugate. 4 3 2 1 is a self-conjugate partition.

The idea of reading a Ferrers graph by rows or by columns is a useful one and can be used to prove such theorems as - the number of partitions of n with exactly k parts are equinumerous with the number of partitions of n whose largest part is k . Another example where the Ferrers graph is very useful is given in exercise [4.15].

4.6   GRAPHS AND TREES

4.6.1   Basic Concepts

A linear graph (often called just a graph) is a collection of nodes (called points or vertices) together with a collection of relationships (called lines or edges) which describe the connections between the nodes.

A typical graph is shown in figure 4.1.

Fig. 4.1.

We will use the terms points and lines instead of the more mathematical terminology vertices and edges which seem to us less descriptive of the way we think about graphs.

A graph is said to be directed if direction is assigned to the lines, otherwise it is undirected.

A directed graph (called a digraph) is shown in figure 4.2. This

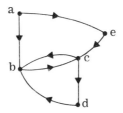

Fig. 4.2. A directed graph

digraph has seven lines (ab), (ae), (bc), (cb), (cd), (db), (ec) and the direction of the line is from the first point to the second.

The graph given earlier, figure 4.1, is an undirected graph and it contains two features which we will often assume are absent from graphs - slings (or loops) where a point is connected to itself by a line as at point F and multiconnections such as the three lines joining B to C.

A graph can be represented $G = (P, L)$ where P is a set of points representing the nodes and L is a set of lines representing the relationships.

For figure 4.2

$$P = \{ a, b, c, d, e \}$$

$$L = \{ (ab), (ae), (bc), (cb), (cd), (db), (ec) \}$$

A path in a graph G is a sequence of lines $(a_1 a_2), (a_2 a_3), (a_3 a_4),$ $\ldots, (a_{n-1} a_n)$ where no line appears more than once in the sequence and the points $a_1, a_2, \ldots, a_n$ all belong to G. If, in addition, the same point is not used twice then the path is an elementary path (or chain). When the starting point a_1 and the finishing point a_n are the same point the path is closed. A closed path is also called a cycle or circuit. An Euler path is a path which traverses each line in the graph exactly once, and an Euler circuit is an Euler path whose start and finish points are the same. A graph is connected if every pair of points is joined by an elementary path. Thus the graph of figure 4.1 is not connected. The idea of connectedness is an important one in graph theory and leads to the definition of trees which are the principal subclass of graphs.

A <u>tree</u> is defined to be a connected linear graph which contains no cycles.

Given a graph G=(P, L) then G'=(P', L') is a <u>subgraph</u> if P' is a subset of P and L' is a subset of L. The <u>complement</u> G"=(P", L") of the subgraph G' with respect to the graph G is another subgraph of G with L"=L-L' and P" contains only points with which the lines of L" are incident. See figure 4.3 below for an example.

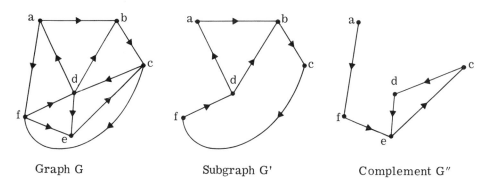

Graph G   Subgraph G'   Complement G"

Fig. 4.3.

An important special case is the <u>spanning tree</u> of a connected graph G ; this is defined to be a tree which is a subgraph of G and which contains all the points of G . For example, figure 4.4 shows an undirected graph G and two of its spanning trees.

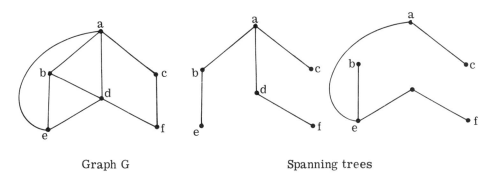

Graph G   Spanning trees

Fig. 4.4.

4.6.2 **Paths in Graphs**

A path in a graph G has been defined as a sequence of lines $(a_1a_2), (a_2a_3), (a_3a_4), \ldots, (a_{n-1}a_n)$ in which no line appears more than once. The <u>length</u> of a path is the number of lines it contains. The <u>degree</u> of a point in a graph is the number of lines which meet at that point.

There are some interesting problems concerning paths in a graph: one of the earliest was ₋ₑe Königsberg bridges problem investigated by Euler in 1736. A map of Königsberg, the river Pregal and the seven bridges is shown in figure 4.5(a). Euler proved it was impossible to

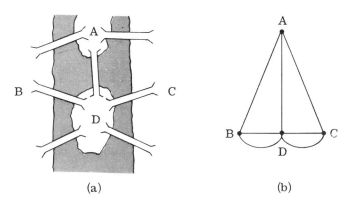

(a)   (b)

Fig. 4.5. The Königsberg bridges and their graphical representation

cross each of the seven bridges once and only once. If the problem is represented as a graph in which the points are the land and the lines are the bridges the graph shown in figure 4.5(b) is obtained. The Königsberg problem is thus equivalent to finding whether the graph has any Euler path. The proof of the basic theorem on Euler circuits requires the following lemma and its important corollary.

Lemma

The sum of the degrees of the points in a graph G is an even number.

Proof. Each line of a graph G has a point at each end and so it adds one to the degree of the two points which are its ends. Therefore the sum of the degrees of the points is twice the number of lines in G.

Corollary. In any graph G there are an even number of points of odd degree.

Basic Theorem on Euler Circuits

An undirected graph possesses an Euler circuit if and only if it is connected and its points are all of even degree.

An outline of the proof proceeds as follows.

The necessity of the conditions is shown by assuming that the graph G has an Euler circuit. In this case, G is connected since there is a path between any two points (i. e. the Euler circuit). When the Euler circuit is traced each time it passes through a point the circuit traverses two lines which meet at that point and which have not been traced before. This shows that all the points except the starting point are of even degree. However the starting point also is of even degree because the line on which the trace starts is different from the line on which it finishes while any other passage through this point adds two to the degree.

The sufficiency of the conditions needs a longer argument. Assume that G is connected and that every point is of even degree. We shall use induction on the number of lines in G . The theorem is obviously true for graphs with one or two lines. Suppose that the theorem is true for all graphs of less lines than those in G . Now start at any point of G and construct a path which finishes at the same point. This can be done because G is connected and each point is of even degree so that we can enter any point on a line and can leave it on a different line. If all the lines of the graph G have been traced we have an Euler circuit and the proof is complete; if not, remove from G all the lines on the circuit C we have just constructed. The resulting graph H , which may be disconnected, has less lines than G and still has all its points of even degree. By the assumption each disconnected component of H has an Eulerian circuit; further each component of H has at least one point on the original circuit C , since G is connected, and so an Euler circuit of G can be obtained by tracing C until a point in H is reached and then tracing the Eulerian circuit of this component before returning to the circuit C . The theorem is thus extended to all graphs with numbers of lines less than, or equal, to those in G , and hence to graphs generally.

Corollary. An undirected graph possesses an Euler path if and only if it is connected and has either exactly two points of odd degree or no such points.

We can now see that in the Königsberg bridges problem the graph in figure 4. 5(b) has all four points of odd degree and so has no Eulerian path.

The ideas in this section can be extended to directed graphs if the incoming and outgoing degrees of a point in a digraph are defined as the number of lines going into and out of the point respectively. A digraph possesses an Euler circuit if and only if it is connected and is such that the incoming degree of every point is equal to its outgoing degree.

A similar but more difficult problem concerns <u>Hamiltonian paths</u>, which are paths passing through each point in a graph exactly once. These are very interesting paths and circuits as they can be generalised into the well-known 'travelling salesman' problem. In this problem a salesman is required to visit a number of towns during a tour, so that he visits each town once and only once and returns to his starting point by the shortest possible route. There is no efficient general algorithm for this problem although much research and computer time have been expended on it (see section 6. 4).

4. 6. 3 Representations of Graphs

The pictorial representation of graphs which has been shown is often convenient for a theoretical study of graphs. However other representations are necessary for computer processing. The simplest method is probably to represent all the lines of the graph by a pair of integers corresponding to its end points. This would certainly be a useful method of input into a computer but is not so satisfactory for some investigations by computer program. For example it would be difficult to test the connectivity of a graph with such a representation. Two useful methods of representing graphs use matrices and they are examined in this section.

The <u>incidence matrix</u> $[I_{ij}]$ represents a graph by a matrix of zeros and ones in which the general element I_{ij} is defined as follows:

$I_{ij} = 1$ if the ith point belongs to the jth line

$\phantom{I_{ij}} = 0$ otherwise

In previous graphs only the points have been labelled but the incidence matrix implies a labelling of the lines as well. Consider the graph in figure 4.6 with its incidence matrix. The order of the columns effectively labels the lines.

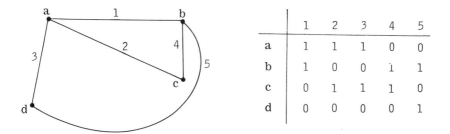

	1	2	3	4	5
a	1	1	1	0	0
b	1	0	0	1	1
c	0	1	1	1	0
d	0	0	0	0	1

Fig. 4.6. A graph and its incidence matrix

The incidence matrix is in general rectangular, and would only be square if the number of lines equalled the number of points. There will be exactly two ones in each column while the number of ones in a row is equal to the degree of the corresponding point. If a graph G is disconnected and consists of components $G_1, G_2, G_3, \ldots$, then the incidence matrix can be partitioned

$$I(G) = \begin{pmatrix} I(G_1) & 0 & 0 & 0 & \cdots \\ 0 & I(G_2) & 0 & 0 & \cdots \\ 0 & 0 & I(G_3) & 0 & \cdots \\ 0 & 0 & \cdot & \cdot & \\ \cdot & \cdot & \cdot & \cdot & \cdot \\ \cdot & \cdot & \cdot & \cdot & \cdot \\ \cdot & \cdot & \cdot & \cdot & \cdot \end{pmatrix}$$

where $I(G_i)$ is the incidence matrix of the graph G_i , and the zero matrices are of the appropriate dimensions.

Incidence matrices are often particularly useful when working with logical or electrical networks.

The <u>adjacency</u> or <u>connection matrix</u> A_{ij} represents a graph as follows:

$A_{ij} = 1$ if the points i and j are joined by a line

$= 0$ otherwise

For the graph given in figure 4.6 the adjacency matrix is

	a	b	c	d
a	0	1	1	1
b	1	0	1	1
c	1	1	0	0
d	1	1	0	0

An adjacency matrix is $(n \times n)$ where n is the number of points in the graph. The leading diagonal is zero if there are no slings in the graph; the adjacency matrix is not usually used to represent a multiconnected graph but the representation could be extended to do so by letting A_{ij} be the number of lines from i to j. The adjacency matrix of an undirected graph is symmetric. If a graph is disconnected the adjacency matrix can be partitioned in a similar way to the incidence matrix.

A further useful property of the adjacency matrix is that the (i, j) element of its pth power, $(A^p)_{ij}$, indicates the number of distinct ways of moving from i to j in just p steps. For example for the matrix above

$$A^3 = \begin{pmatrix} 4 & 5 & 5 & 5 \\ 5 & 4 & 5 & 5 \\ 5 & 5 & 2 & 2 \\ 5 & 5 & 2 & 2 \end{pmatrix}$$

Thus there are only two ways of getting from d to c in three steps (which inspection shows to be $d \to a \to b \to c$, or $d \to b \to a \to c$).

Another representation useful for some purposes is to hold for each point a list of the points to which it is directly connected by a line. For the example in figure 4.6 the list would be:

a: b, c, d
b: a, c, d
c: a, b
d: a, b

This representation will be called an adjacency structure and has obvious connections with the adjacency matrix. In fact it is the adjacency matrix represented as a sparse matrix. Both the adjacency matrix and the adjacency structure show a certain degree of redundancy when they are used to represent undirected graphs but the structure can be convenient for finding paths through graphs when the internal computer representation might employ linked lists.

4.6.4 Shortest Paths in Graphs

The notation and ideas introduced in the previous sections are illustrated by applying them to the problem of finding the shortest distance between points in a graph. The graph G to be considered has n points and a distance associated with each line of the graph. (Such a graph is often called a network.) The representation of the network will be as a distance matrix D which has non-zero elements in the same positions as the adjacency matrix except that instead of zeros and ones the actual distances between the points are used. If the points i and j are not joined then a distance much greater than any possible path length in the network is inserted in the network. Let the distance matrix be $D = (d_{ij})$ where

$$d_{ij} = 0 \quad \text{if } i = j$$

$$= \infty \quad \text{if } i \text{ is not joined to } j \text{ by a line}$$

$$= \text{distance of the line from } i \text{ to } j \ (> 0)$$

At any stage during the method to be described there will be two sets of points K and U, where K consists of those points which have been fully investigated and between which the best path is known, and U of those points which have not yet been processed. Clearly every point belongs to either K or U but not to both. Let a point r be selected from which we shall find the shortest paths to all the other points of the network. Let the array bestd(i) hold the length of the shortest path so far formed from r to point i, and another array tree(i) the next point to i on the current shortest path. The algorithm usually known by Dijkstra's name proceeds as follows:

Step 1. Initialise : K {point r }, U {all the other points except r }. For all nodes except r set $bestd(i) = d_{ri}$ and $tree(i) = r$.

Step 2. Find the point s in U which has the minimum value of bestd . Remove s from U and put it in K .

Step 3. For each node u in U find $bestd(s) + d_{su}$ and if it is less than bestd(u) replace bestd(u) by this new value and let $tree(u) = s$. (In other words a shorter path to u has been found by going via point s .)

Step 4. If U contains only one point stop else go back to step 2.

When the algorithm terminates bestd(i) contains the length of the shortest path from r to i and that path is $i \rightarrow tree(i) \rightarrow tree(tree(i)) \ldots$ until r is reached. A detailed algorithm, using the same notation as above, is given below.

```
procedure Minpath (integer value n, r; integer array
                dist (*, *); integer array bestd, tree(*));
begin integer p, s, min, tempd, u; integer array
    Unknown (1::n);
    for i: = 1 until n do
    begin Unknown(i):=i; bestd(i):=dist(r, i);
            tree(i):=r
    end;
    Unknown(r):=n;
    comment r is not in set Unknown so replace it by n;
    for k:=1 until n-2 do
    begin p:=1; min:=bestd(1);
            for i:=2 until n-k do
            if bestd (Unknown(i)) < min
            then begin p:=i; min:=bestd(Unknown(i))
                    end;
```

```
        s:=Unknown(p);Unknown(p):=Unknown(n-k);
        for i:=1 until n-k-1 do
        begin u:=Unknown(i);
                tempd:=min+dist(s, u);
                if bestd(u) > tempd then
                begin bestd(u):=tempd;tree(u):=s
                end
        end
    end k loop;
end Minpath;
```

Let us now see how Dijkstra's algorithm works through the following example.

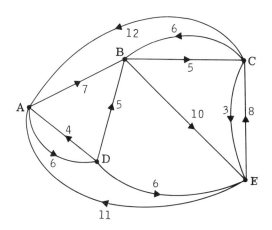

Fig. 4. 7.   Network with distances on the lines

Distance matrix

	A	B	C	D	E
A	0	7	∞	6	∞
B	∞	0	5	∞	10
C	12	6	0	∞	3
D	4	5	∞	0	6
E	11	∞	8	∞	0

Let the selected node be B . In step 1 we have K{B} and U{A, C, D, E},
with the arrays

	A	C	D	E
bestd	∞	5	∞	10
tree	B	B	B	B

1st iteration. Minimum (bestd)=5 . Remove node C from U and put it
in K . Test to see if it is shorter to go from B to A, D or E via C ;
it is to A since $17 < \infty$. Put bestd(A)=17, tree(A)=C . Also for E
$5+3 < 10$ so put bestd(E)=8, tree(E)=C .

2nd iteration. Minimum (bestd)=8 . Remove node E from U and put it
in K . Test to see if it is shorter via E to A or D . It is not.

3rd iteration. Minimum (bestd)=17 . Remove node A from U and put
it in K . It is shorter to node D via node A so we set bestd(D)=23,
tree(D)=A .

There is now only one point D in set U so the algorithm ends. The final
arrays are

	A	C	D	E
bestd	17	5	23	8
tree	C	B	A	C

4.6.5 Trees

In this section we examine briefly the important subclass of graphs
called trees, which have already been defined as connected linear graphs
with no circuits. Two properties of trees which follow from this definition
are:

(i) Any two points in a tree are connected by a unique path.

(ii) A tree with n points has n-1 lines.

We can prove (i) immediately since there must be a path between
two points in a connected graph and if there were more than one path there
would be a circuit which is impossible by the definition of a tree. The

property (ii) is more difficult to prove but can be achieved by induction on n , the number of points. Clearly a tree with two points has one line. Assume (ii) is true for all trees with less than n points. A tree must have at least one point of degree one, since a point of degree zero would give a disconnected graph and if all points had degree greater than one there would be a circuit. Remove the point of degree one and the line connected to it; this leaves a tree with n-1 points which by the induction hypothesis has n-2 lines. Therefore adding back the point and line we have removed gives a tree with n points and n-1 lines.

The trees we shall consider in this section will be <u>rooted trees</u> in which a special node is designated as the root. If the tree is a directed tree the root has no incoming lines and the incoming degree of all the other points is one. Some of the notation and ideas used in trees are taken from family trees, those diagrams used by genealogists to indicate one's ancestors and descendants. Accordingly the trees will be drawn here top down as shown below.

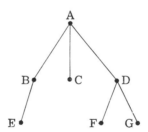

4.6.6 A Spanning Tree Algorithm

A spanning tree of a graph G as defined in section 4.6.1 is a tree which contains all the points of G . Algorithms for finding spanning trees are some of the most frequently used in graph theory. There are many variations and we will only look at one of the simplest problems, i.e. that of finding one spanning tree. Finding all the spanning trees is considerably more difficult, and another related problem is finding the minimum spanning tree, which assumes that the lines of the graph have weights associated with them so that the spanning tree whose lines have the minimum total weight is required.

Suppose G is an undirected graph with n points and m lines and no slings; let its n points be numbered $1, 2, 3, \ldots, n$ and let G be represented by two integer arrays A and B where, for each line, A contains the first point of the line and B the other point. Therefore $A(i), B(i)$ are the end points of the ith line.

Initially the spanning tree is empty and at each stage of the algorithm the next line is tested to see if either or both its end points appear in the spanning trees formed so far.

At the ith stage $1 \le i \le m$ when we examine the line $A(i), B(i)$ we must recognise the following conditions:

(a)   Neither $A(i)$ nor $B(i)$ is in any of the trees so far constructed. The ith line therefore becomes a new tree with component number c, $c:=c+1$.

(b)   Point $A(i)$ is in tree T_j and $B(i)$ in tree T_k (where $j \neq k$, $1 \le j$, $k \le c$). The ith line is used to join these two trees, the points of T_k are given the component number of T_j . $c:=c-1$.

(c)   Both points are in the same tree; ignore this line.

(d)   Point $A(i)$ is in tree T_j and $B(i)$ is not in any tree. The ith line is added to T_j and $B(i)$ is given the component number of T_j .

(e)   Point $A(i)$ is not in any tree and $B(i)$ is in tree T_k . The ith line is added to T_k and $A(i)$ is given the component number of T_k .

The efficiency of the algorithm depends mainly on the speed with which we can test whether or not the end points of the lines are in the spanning trees. In order to do this testing efficiently an integer array Point(1::n) is used. This array has $Point(A(i)) = Point(B(i)) = c$ if the ith line (with end points $A(i), B(i)$) belongs to the cth spanning tree. Thus if $Point(X) = 0$ then X has not yet been used in any spanning tree. We also need an integer array LINE(1::m) which indicates which line belongs to which spanning tree. Thus if $LINE(k) = c$ then the kth line belongs to the cth spanning tree and if $LINE(k) = 0$ then the line does not belong to any spanning tree. A detailed algorithm is given below.

```
procedure SPANNING TREE (integer array A, B, LINE(*);
                        integer value m, n);
  begin integer r, s, c;  integer array Point(1::n);
        c:=1;
        for i:=1 until n do Point(i):=0;
        for j:=1 until m do
        begin r:=Point(A(j)); s:=Point(B(j));
          if r=0 and s=0 then
             begin LINE(j):=Point(A(j)):=Point(B(j)):=c;
                c:=c+1; end
             else if r=0 and s≠0 then
             LINE(j):=Point(A(j)):=s
             else if r≠0 and s=0 then
             LINE(j):=Point(B(j)):=r
             else if r≠s then
             begin LINE(j):=r;
                   for k:=1 until n do
                   if Point(k):=s then Point(k):=r;
          end
        end j;
  end SPANNING TREE;
```

The algorithm will give one spanning tree if the graph is connected and one spanning tree for each connected component if it is not connected. It can therefore be used to test whether or not a graph is connected.

If we take as an example the graph G in figure 4.4, then the spanning tree found by the algorithm will depend on the order in which the lines are input. If the order is (ab), (ac), (ad), (bd), (be), (de), (df), (ea), (fc) then the first of the spanning trees shown in figure 4.4 is obtained.

4.7 BIBLIOGRAPHY

The mathematical properties of permutations, combinations, partitions and compositions are covered in many textbooks. Although it is now twenty years old the book by Riordan [1] is still one of the best; two other general books which can be well recommended are by Liu [2] and Hall [3]. All three books treat generating functions, of which another view may be seen in Knuth [4] and yet another in Feller [5] who considers applications to probability. The theory of enumeration and generating functions is extensively covered in the classical work of MacMahon [6] which has relatively recently been reprinted, but the detailed treatment in this book's rather dated notation makes it less suitable for the casual reader.

The number of mathematical textbooks on graphs is even larger. The introduction to graphs in Liu's book and the section on trees in Riordan are both good. Berge [7] is one of the best of the older textbooks on graphs and from the plethora of recent ones we would select Harary [8], Deo [9], and Wilson [10]. An introduction to the computer representation of trees with several associated algorithms is given by the authors in [11] and this is discussed in greater detail in Knuth [4]. The shortest path algorithm is adapted from Dijkstra's method [12] and the spanning tree algorithm from that of Seppänen [13].

Finally, the book by Wells [14] covers the material of this chapter and many others in this book. He gives many interesting results, representations and algorithms but we believe that someone starting the subject may not find it an easy book to read.

[1]   J. Riordan: 'An Introduction to Combinatorial Analysis.' John Wiley, 1958.

[2]   C.L. Liu: 'Introduction to Combinatorial Mathematics.' McGraw-Hill, 1968.

[3]   Marshall Hall, Jr.: 'Combinatorial Theory.' Blaisdell Pub. Co., 1967.

[4]   D.E. Knuth: 'The Art of Computer Programming. Vol 1. Fundamental Algorithms.' pp 44-9. Addison-Wesley, 1968.

[5]   W. Feller: 'An Introduction to Probability Theory and its Applications. Vol 1.' 3rd Edition, Chapter XI, pp 264-85. John Wiley, 1968.

[6]   P.A. MacMahon: 'Combinatorial Analysis.' Vols I and II (first published 1915, 1916) reprinted in one volume. Chelsea Pub. Co., 1960.

[7]   C. Berge: 'The Theory of Graphs and its Applications.' John Wiley, 1962.

[8]   F. Harary: 'Graph Theory. ' Addison-Wesley, 1969.

[9]   N. Deo: 'Graph Theory with Applications to Engineering and Computer Science. ' Prentice-Hall Inc. , 1974.

[10] R. J. Wilson: 'Introduction to Graph Theory. ' Oliver and Boyd, 1972.

[11] E. S. Page and L. B. Wilson: 'Information Representation and Manipulation in a Computer. ' Cambridge University Press, 1973. Chapter 6.

[12] E. W. Dijkstra: 'A Note on Two Problems in Connection with Graphs. ' Numerische Math. , Vol 1 (1959), pp 269-71.

[13] J. J. Seppänen: 'Algorithm 399: Spanning Tree. ' Comm. ACM, Vol 13, No 10 (1970), pp 621-2.

[14] M. B. Wells: 'Elements of Combinatorial Computing. ' Pergamon Press, 1971.

4. 8   EXERCISES

[4. 1]   How many ways may one right and one left shoe be selected from six pairs of shoes without obtaining a pair ?

[4. 2]   If five men A, B, C, D, E intend to speak at a meeting, in how many orders can they do so without B speaking before A ? How many orders are there in which A speaks immediately before B ?

[4. 3]   How many ways can twelve white pawns and twelve black pawns be placed on the black squares of an 8×8 chess board?

[4. 4]   How many positive integers less than one million can be formed using sevens, eights, nines only? How many using zeros, eights and nines only

[4. 5]   Find the sum of all the four digit numbers that can be obtained by using the digits 1, 2, 3, 4 once in each.

[4. 6]   How many points of intersection are formed by n lines drawn in a plane if no two are parallel and no three concurrent? Into how many regions is the plane divided?

[4. 7]   How many ways can three integers be selected from $3n$ consecutive integers so that the sum is a multiple of 3 ?

[4. 8]   How many ways can five different messages be delivered by three messengers if no messenger is left unemployed? The order in which a messenger delivers his messages is immaterial.

[4. 9]   In how many ways can a lady wear five rings on the fingers (not the thumb) of her right hand?

[4. 10] Six distinct symbols are transmitted through a communication channel. A total of twelve blanks are to be inserted between the symbols with at least two blanks between every pair of symbols. In how many ways can we arrange the symbols and blanks?

[4. 11] A new national flag is to be designed with six vertical stripes in yellow, green, blue and red. In how many ways can this be done so that no two adjacent stripes have the same colour?

[4. 12] The Fibonacci sequence of numbers $0, 1, 1, 2, 3, 5, 8, 13, 21, 34, \ldots$ is such that each number is the sum of the two preceding numbers. Derive a generating function

$$G(z) = F_0 + F_1 z + F_2 z^2 + \ldots + F_n z^n + \ldots$$

for the Fibonacci sequence, and use it to show that

$$F_n = \frac{1}{\sqrt{5}} (\alpha^n - \bar{\alpha}^n)$$

where $\alpha = (1+\sqrt{5})/2$ and $\bar{\alpha} = (1-\sqrt{5})/2$

[4. 13] How many compositions of n with m parts are there when zero parts are allowed?

[4. 14] Prove that the number of partitions of n in which no integer occurs more than twice as a part is equal to the number of partitions of n into parts not divisible by 3. Thus if $n = 6$ the partitions in which no integer occurs more than twice are 6, 51, 42, 411, 33, 321, 2211. The partitions in which no part is divisible by 3 are 51, 42, 411, 222, 2211, 21^4, 1^6.

[4.15] Show that the number of partitions of n is equal to the number of partitions of $2n$ which have exactly n parts.

[4.16] How many ways can an examiner assign 30 marks to 8 questions so that no question receives less than 2 marks?

[4.17] Show that the number of partitions of n in which every part is odd is equal to the number of partitions of n with unequal (or distinct) parts.

[4.18] Prove that the number of combinations of n distinct objects taken m at a time with unrestricted repetition is

$$\binom{n+m-1}{m}$$

What are these combinations for $m = 3$ and the four objects w, x, y and z?

A shop sells six different flavours of ice-cream. In how many ways can a customer choose four ice-cream cones if:
(a) they are all of different flavours;
(b) they are not necessarily of different flavours;
(c) they contain only two or three flavours;
(d) they contain three different flavours?

(Newcastle 1968)

[4.19] Consider:
(a) The partitions of n into three parts

$$n = n_1 + n_2 + n_3$$

where $n_1 \geq n_2 \geq n_3$
and

(b) The partitions of $2n$ into three parts such that the sum of any two parts is greater than the third

$$2n = p_1 + p_2 + p_3$$

where $p_1 \geq p_2 \geq p_3$ and $p_i + p_j > p_k$ $(i \neq j \neq k)$.
Using the construction

$$p_1 = n_1 + n_2 = n - n_3$$
$$p_2 = n_3 + n_1 = n - n_2$$
$$p_3 = n_2 + n_3 = n - n_1$$

show that from every partition of (a) we can obtain one of (b) and vice versa. Show further that the partitions (a) and (b) are equinumerous.

(Newcastle 1970)

[4.20] Given integers $1, 2, 3, \ldots, 11$, two groups are selected; the first group contains five integers and the second group contains two integers. In how many ways can the selection be made with unrestricted repetition if
(a) There are no further restrictions?
(b) A group contains either all odd integers or all even integers?

If no repetition is allowed, in how many ways can the selection be made such that the smallest member of the first group is larger than the largest member of the second group?

(Newcastle 1971)

[4.21] If a number n has only two different prime factors p_1 and p_2 show that $f(n)$, the number of positive integers less than n and prime to n, is given by

$$f(n) = n(1 - \frac{1}{p_1})(1 - \frac{1}{p_2})$$

What are $f(12)$ and $f(135)$?

(Newcastle 1966)

[4.22] (i) Use the principle of inclusion and exclusion to find the number of ways to distribute r distinct objects into n distinct cells so that no cell is empty.

(ii) Solve the same problem using exponential generating functions to count r-permutations of n objects with repetitions and each object occurring at least once.

(iii) Verify that the result $T(r, n)$ satisfies the recurrence relation:

$$T(r, n) = n(T(r-1, n) + T(r-1, n-1))$$

(Warwick 1975)

[4.23] (i) Prove the principle of inclusion and exclusion.

(ii) Hence show that the proportion of permutations of $\{1, \ldots, n\}$ which contain no consecutive pair i, $i+1$ for any i is approximately $(n+1)/en$.

(Warwick 1975)

[4.24] State and prove the principle of inclusion and exclusion. Hence find the number of derangements of the symbols $\{1, 2, \ldots, n\}$. What proportion of all permutations of $\{1, 2, \ldots, n\}$ does this represent if n is large?

(Warwick 1976)

[4.25] Define and give examples of the following:

(a) The Ferrers graph;

(b) Conjugate partitions;

(c) A self-conjugate partition.

Prove that the number of partitions of n with exactly k parts equals the number of partitions of $n-k$ with no part greater than k . Show what these partitions are for $n = 11$, $k = 3$.

(Newcastle 1967)

[4.26] Let p_n be the number of unrestricted partitions of n , and p_n^* be the number of partitions of n without unit parts (e.g. $p_4^* = 2$, the partitions are 4 and $2\,2$).

What are the partitions corresponding to p_{10}^* ?

94

Using generating functions or otherwise show that, for $n > 1$, $p_n^* = p_n - p_{n-1}$. Generalise this result to find a formula for the number of partitions of n without parts of size k.

Also show that $p_{n+2} - 2p_{n+1} + p_n \geq 0$.

(Newcastle 1969)

[4.27] $p(t, s, k)$ represents the generating function for partitions with exactly s parts and with greatest part equal to k.

Explain what this means and show that

$$p(t, s, k) = p(t, k, s)$$

Show that

(a) $p(t, 1, k) = t^k$

(b) $p(t, 2, k) = t^{k+1}(1-t^k)/(1-t)$

(c) $p(t, s, k) = t^k p(t, s-1, k) + tp(t, s, k-1)$

Hence, or otherwise, show that

$$p(t, s, k) = \frac{t^{s+k-1}J(t, s+k-2)}{J(t, k-1) \times J(t, s-1)}$$

where

$$J(t, n) \equiv (1-t^n)(1-t^{n-1}) \ldots (1-t^2)(1-t)$$

Write down explicitly the partitions of 11 with 3 parts, greatest part 5 and relate their number to $p(t, 3, 5)$.

(Newcastle 1972)

[4.28] In the Königsberg bridges problem what is the minimum number of bridges that have to be removed to make the graph in figure 4.5(b) have an Euler circuit? Which bridge or bridges would you remove?

[4.29] Prove that a connected graph G remains connected after removing the line L from G if and only if L is in some circuit in G.

[4. 30] What type of graph becomes disconnected when any line is removed from it?

[4. 31] An undirected graph G has n points and m lines and is specified on data cards by the two integers n and m followed by m pairs of numbers representing the lines of G . Two points P and Q in the graph are specified.

Write an ALGOL W program (or detailed algorithm or detailed flow diagram) which will determine if there is an Euler path from P to Q .
If there is no Euler path this should be stated on the output together with the set of lines whose addition to G would give an Euler path from P to Q .

Describe what results your algorithm would give for the following data:

$n = 5$, $m = 8$
lines: $(1, 2)$, $(1, 3)$, $(1, 4)$, $(1, 5)$, $(2, 3)$, $(2, 4)$, $(3, 4)$, $(4, 5)$.
$P = 1$, $Q = 3$.

(Newcastle 1973)

[4. 32] A directed graph G has n points and m lines and is specified on data cards by two integers n and m followed by m pairs of numbers representing the lines of G . (The direction of the line is from the first number of the pair to the second.) Two points P and Q on the graph are specified. Write an ALGOL W program which will determine whether there is a path from P to Q . This program should be efficient and can either use arrays or RECORDS and REFERENCES.

5 · Ordering and Generation of Elementary Configurations

5.1 LEXICOGRAPHICAL ORDERINGS

5.1.1 Filing Rules

The idea of an alphabetical order is a familiar one. It is used in dictionaries, book and card indexes, directories and many other works of reference. In each of these reference tools a collection of letters in a particular order is a key which determines the place it will appear in the reference work together with some accompanying other information, e. g. in a dictionary the key is a word of the language, and the information which is associated with it is its definition, or its synonyms in a foreign language, and perhaps illustrations of the use of the key word in literature or notes on its linguistic derivation. The fundamental idea is that of a key, which is used to fix the position of the entry in the order, and the information associated with that key. Thus the normal telephone directory lists names of subscribers in alphabetical order and the associated information is their telephone numbers and their addresses, the latter perhaps in an abridged form. But even though the idea of an alphabetical order is so familiar many compilers of such reference works find it necessary to explain to their users some parts of what system of ordering they have used. For example, many persons with Celtic ancestors have family names which begin Mac, Mc or M^c; these similar prefixes are evidently a source of confusion to those wishing to make a telephone call to a person with a name of that form and the compiler of the directory may take pity on the caller by making the decision to treat all of these prefixes as being the same for the purposes of ordering and even to place them in some prominent position, say at the beginning of the list of those whose names begin with M. This is adopting what librarians would call a 'filing rule'. The compiler chooses such rules for what he judges to be the convenience of the users of his reference tool. Usually there will be several, even many, decisions of this kind to be made. Is 'IBM', that universally used abbreviation for the International Business Machines Company, to be filed as a word beginning with Ib and so precede the oriental name Ibrahim, or as its unabbreviated form which will be filed at a later position?

Even keys composed solely of letters of the same alphabet can thus give rise to some questions of precedence. Still others appear when numbers can occur as well. Should the 'A1 Manufacturing Company' appear before a man named Aaron, or after him? Librarians of scholarly collections are likely to have a variety of alphabets often mixed together in the titles they have to file. If the ordering is to be a precisely defined one, it is clear that there must be rules to cover all the classes of cases that will be encountered in order to allow the list to be constructed at all, or to receive additions in a non-contradictory manner, never mind about it being able to be used subsequently. Next, for the list to be used, the rules must be known, be available for reference or at least susceptible of being guessed with little difficulty. These are the minimum requirements for a list to be formed and to be used, but they say nothing about how the formation of the rules should be influenced by how the ordering is to be used. An English speaking scientist who wishes to use his imperfect knowledge of the French language to translate a paper from that language into his own tongue can use a suitable French-to-English dictionary to help him. An English-to-French dictionary would be almost no use for this purpose. One of the well-known English dictionaries is perfectly suited for obtaining the meaning of an abstruse word encountered in reading; it is not quite so good for establishing the spelling of words which are imperfectly known. For example, uncertainty about the correct number of r's and s's in a word like 'embarrassment' will be quite quickly established because the filing positions of the conceivable alternatives are close together. This is not the case for words which might only have been overheard; a search for the name of the flower 'phlox' would be abortive if one concentrated one's attention near the plural of a word for a group of sheep, or for some wool wadding used in inferior pillows. Such a dictionary is also not very well suited for supplying words of similar or related meaning - for which a thesaurus-type of construction is more useful. Rhymesters and crossword puzzle addicts might like a listing in alphabetical order of the endings of words rather than the beginnings. Those interested in linguistics might use a variety of other listings.

We use the term <u>systematic ordering</u> for any ordering of a set of items, whether they be words of a single alphabet or of mixtures of several alphabets, numbers, or other symbols, produced in accordance with a set of unambiguous rules. The reader may encounter the term <u>lexicographical ordering</u> in other literature where we would use 'systematic ordering'; we shall restrict the adjective 'lexicographical' to those orderings which are alphabetical in the usual sense, or simple extensions of the alphabetic notation - for example, if the individual symbols of the key are the digits then the lexicographical ordering would show the keys in increasing numerical order. The examples above have illustrated some of the operations that might need to be performed; some more are made explicit in the list which follows.

1. Given a set of items, produce a systematic ordering;
2. Given a new item, insert it into its correct place in the ordering;
3. Given an item in the ordering, find its position, i. e. find how many places it is removed from the position of the first item;
4. Given a position, determine the item in the ordering which occupies that position;
5. Given an element in the ordering, generate the next element;
6. Generate a random member of the list.

The ease with which these operations may be performed will depend upon a variety of factors - whether the whole list has been stored, how it has been stored, the facilities available in the high-level language if one is being used, and the order structure, architecture and speed of the machine. As usual in computing, the design of a system or piece of program should take into account the frequency with which different operations have to be performed. Some of the operations listed above are just the familiar ones of sorting and searching about which much has been written, and algorithms for them will have been included in other courses in computing science.

5.1.2 Number Systems

In order to obtain a systematic order for a set of items it is enough to have unambiguous, consistent and complete rules for determining which of a pair of items should precede the other for all possible pairs. This is the situation when one has filing rules referred to in the previous sub-section. If the knowledge of the set and the way the ordering is formed is limited in this way, the problems of identifying where a given set member occurs in the list or which member is at a specified place will need to be determined by referring to the list itself, perhaps using one of the standard search algorithms. In some cases, however, more information will be available. The number, n , of items in the set may be known and more particularly it may be possible to place the items in one-to-one correspondence with the integers $1, 2, \ldots, n$ by an algorithm without using the whole list. In terms of searching algorithms this last requirement is equivalent to there being a rule for pigeon-holing the items which is derived from the method of establishing the one-to-one correspondence; an 'address calculation' algorithm is available which is exact.

In some problems, different methods of representing the integers can be useful; some will already be familiar. For example, a natural systematic order for the squares on a standard chess board would be to treat the cells row by row from left to right so that the jth cell in the ith row would correspond to the integer $8(i-1) + j$. The generalisation to a $b \times b$ sized board or to a board of k dimensions is straightforward - the formula establishing the correspondence is similar to that giving the address at which the elements of a square k-dimensional array are stored normally in sequential storage and is of the form

$$\sum_{i=1}^{k} (a_i - 1)b^{i-1} + 1 \tag{5.1}$$

where $1 \le a_i \le b$, which is clearly an expression for an integer in the range 1 to b^k in the scale of b . For some purposes this integer is most conveniently identified by the coordinates $(a_1, a_2, \ldots, a_k)$ while for others it is better to use the integer value given by (5.1) for which $a_1, a_2, \ldots, a_k$ are simply related to the digits in the b-ary scale. In some other problems less familiar types of number representation can be useful.

One such representation is the expression of the integers in terms of 'factorial digits'.

The simplest factorial representation of the integers is

$$m = a_1 1! + a_2 2! + a_3 3! + \ldots + a_{n-1}(n-1)! \tag{5.2}$$

where $0 \le a_i \le i$; this representation is easily shown to be unique. The a_i's are called the factorial digits of m and we can write

$$m = (a_1, a_2, \ldots, a_{n-1})$$

For example

$$0 = (0, 0, \ldots, 0)$$
$$2000 = (0, 1, 1, 3, 4, 2)$$

Conversion from the decimal representation to factorial representation is obtained by dividing m by 2 , giving the remainder a_1 , then dividing the quotient by 3 , giving the remainder a_2 , etc. Thus we let

$$m_0 = m$$

$$m_i = [m_{i-1}/(i+1)] \text{ for } i > 0, \text{ where } [x] \text{ is the whole number part.}$$

Then

$$a_i = m_{i-1} - (i+1)m_i \tag{5.3}$$

An alternative derivation of the a_i is obtained by successively dividing by i! , which obtains a_{n-1} first, then a_{n-2} , and finally a_1 .

The factorial representation using the (n-1) digits $a_1, a_2, \ldots, a_{n-1}$ can represent all the integers from 0 to n!-1 . The largest integer is represented by $(1, 2, 3, \ldots, n-1)$. We can therefore seek to establish a one-to-one correspondence between the n! permutations of n marks and the n! factorial representations $(a_1, a_2, \ldots, a_{n-1})$, and any one-to-one correspondence between the latter and the natural numbers by (5.2) provides a systematic ordering. Not surprisingly, the factorial digits can be used in a variety of ways for ordering permutations of n distinct objects. A typical example is the derangement method of Marshall Hall where the values of the a_i $(1 \le i \le n-1)$ are taken as the number of marks less than i which actually follow i in the permutation. This definition ensures $0 \le a_i \le i$ and gives the required one-to-one correspondence between the factorial representations and the permutations.

The above description of the derangement method assumed the marks were $0, 1, 2, \ldots, n-1$. Table 5.1 shows the one-to-one correspondence between the factorial digits and the permutations of the $n = 4$ marks 0, 1, 2, and 3.

Number	Factorial digits			Permutation
	a_1	a_2	a_3	
0	0	0	0	0123
1	1	0	0	1023
2	0	1	0	0213
3	1	1	0	1203
4	0	2	0	2013
5	1	2	0	2103
6	0	0	1	0132
7	1	0	1	1032
8	0	1	1	0231
9	1	1	1	1230
10	0	2	1	2031
11	1	2	1	2130
12	0	0	2	0312
13	1	0	2	1302
14	0	1	2	0321
15	1	1	2	1320
16	0	2	2	2301
17	1	2	2	2310
18	0	0	3	3012
19	1	0	3	3102
20	0	1	3	3021
21	1	1	3	3120
22	0	2	3	3201
23	1	2	3	3210

Table 5.1 One-to-one correspondence between the factorial digits and the permutations used in the derangement method for $n = 4$

An alternative representation which uses factorial digits is defined as follows:

$$m = n!\left(\frac{b_2}{2!} + \frac{b_3}{3!} + \ldots + \frac{b_n}{n!}\right) \tag{5.4}$$

$$= b_n + nb_{n-1} + n(n-1)b_{n-2} + \ldots + n(n-1)\ldots 4.3b_2$$

where $0 \le b_i < i$.

The digits $b_2, b_3, \ldots, b_n$ can be used to represent all the integers from 0 to $n!-1$ and thus we can set up a one-to-one correspondence between the permutations and these digits. This representation is the basis of the adjacent mark method of generating permutations (see section 5.2.3).

5.2   PERMUTATIONS

Permutations are required very frequently in combinatorial and other problems and over the years a large number of methods of generating them by computer have been developed. Before deciding which method to use to generate the permutations we should have a clear idea which of the following requirements is most important.

(a) The generation of all the permutations.

(b) The generation of all favourable permutations.

(c) The generation of random (or representative) permutations.

(d) The generation of permutations satisfying certain conditions.

Looking at these requirements in order, we must remember in (a) that unless the size of the permutation is small the problem will soon become intractable even with the fastest of modern computers. Problems which require the generation of all permutations will get out of hand quickly - note that $5! = 120$, $10! = 3.6 \times 10^6$, $15! = 1.3 \times 10^{12}$, $20! = 2.4 \times 10^{18}$, $25! = 1.55 \times 10^{25}$. It will be obvious from such numbers that it is quite impractical in most cases to store all the permutations; a permutation must be generated, examined, and only in special circumstances retained, e.g. because it is the best one for our problem that we have found so far. The time required is also enormous when larger sizes of problem are reached - if our computer can process permutations at the (quite high) rate of one million per second then for

$n = 20$ the time required for all of them is 2.4×10^{12} seconds $\simeq 80\,000$ years, while even for $n = 15$ it is still of the order of 15 days.

Thus by the time n is about 12 we have to find alternative strategies instead of generating all permutations. One of these is (b), and in this case we try to cut down the number of permutations by skipping over large blocks of unwanted permutations. Consider for example the 'travelling salesman' problem for 12 towns which seeks a shortest tour passing through each of the towns $A, B, C, \ldots, K, L$ once and once only; suppose the towns can be divided geographically into three well-separated clusters as in figure 5.1.

A   B
•   •   •E   •F   •G   •H

C• D•   I•   •J

K•   •L

Fig. 5.1.

It would seem reasonable to ignore all permutations which start with one town from each block as a shortest tour can hardly arise from such a poor start. So we can skip all permutations beginning A I E no matter what the rest of the permutation is. Thus we can ignore $9! = 362\,880$ permutations without generating them and there are other similar reductions possible.

The computer program which generates only those permutations which are worth examining is of course considerably more complicated than the one used to generate all the permutations, but the saving may still be great.

Although the method discussed above may help in some cases, very often the problems are just so big that we can only contemplate the approach (c), generating a few permutations which we hope to chose well so that they are good representatives of the general set of all permutations. Care is, however, needed in some applications of this approach. Suppose we are searching for a counter example to disprove a conjecture about permutations and we consider the case $n = 25$; there might be a tendency to think that after we have tested a few million random permutations without finding a counter example that one is unlikely to occur in the whole set of permutations. This could be quite wrong; we have actually examined only a tiny fraction of the total search space. We must therefore be careful what we conclude with tests carried out using random permutations.

Finally the situation in (d) is somewhat similar to that discussed in (b). The number of marks to be permuted is probably large but only a relatively small number of the permutations are admissible. The method must generate only those permutations which satisfy the restrictions imposed. As in (b) the computer program required is likely to be specialised and quite complex.

A common situation in generating permutations is that we require permutations of m objects drawn from n distinct objects, which sometimes may be repeated in the permutation. An algorithm for this case is given in section 5.2.7.

5.2.1 The Ordering of Permutations

There are many different ways in which the permutations can be ordered and some of the more common ones are given in table 5.2. If the problem requires the generation of all the permutations then the algorithm employed is usually one which derives the next permutation from the previous one. This has the advantage that only the current permutation has to be available for entry into the procedure and it is not necessary to store any auxiliary counters. For any given order we can also seek methods of deriving the ith permutation in that ordering directly, i.e. without generating any of the earlier ones in the order. A permutation will not necessarily occupy that same position in different orderings; for example from table 5.2 the 4th permutation happens to be different for each ordering given. We will examine now the orderings given in table 5.2.

(i)   Lexicographical. This is equivalent to an alphabetical ordering if the marks are letters. Thus ABEL comes before ABLE and so on. In effect the first mark is kept in the first position for as long as possible whilst the other marks are permuted. This type of ordering was discussed in detail in 5.1.1.

(ii)   Reverse lexicographical. Here the last mark is kept in position for as long as possible while the first n-1 marks are changed.

(iii) M. B. Wells' order is given by one of the methods in which a new permutation is obtained from the previous one merely by exchanging two marks. The marks exchanged are usually adjacent but not always (cf. permutations 14, 15).

Permu-tation number	Lexico-graphical	Reverse lexico-graphical	M. B. Wells' order	Adjacent mark order	Fike's order	Repeated objects – lexico-graphical order
1	1234	1234	1234	1234	1234	1111
2	1243	2134	2134	1243	1243	1112
3	1324	1324	2314	1423	1432	1113
4	1342	3124	3214	4123	4231	1114
5	1423	2314	3124	4132	1324	1121
6	1432	3214	1324	1432	1342	1122
7	2134	1243	1342	1342	1423	1123
8	2143	2143	3142	1324	4321	1124
9	2314	1423	3412	3124	3214	1131
10	2341	4123	4312	3142	3241	1132
11	2413	2413	4132	3412	3412	1133
12	2431	4213	1432	4312	4213	1134
13	3124	1342	1423	4321	2134	1141
14	3142	3142	4123	3421	2143	1142
15	3214	1432	2143	3241	2431	1143
16	3241	4132	1243	3214	4132	1144
17	3412	3412	4213	2314	2314	1211
18	3421	4312	2413	2341	2341	1212
19	4123	2341	2431	2431	2431	1213
20	4132	3241	4231	4231	4312	1214
21	4213	2431	3241	4213	3124	1221
22	4231	4231	2341	2413	3142	1222
23	4312	3421	4321	2143	3421	1223
24	4321	4321	3421	2134	4123	.

(256 perms)

Table 5.2   The ordering of permutations (n = 4)

106

(iv) Adjacent mark order is an example of an order in which the next permutation is derived from the previous one always by swapping two adjacent marks. One way of obtaining this ordering is by the following recursive method.

(a) If $n = 2$, the permutations are 1 2 and 2 1 .

(b) If $n > 2$, make n copies of each of the permutations of n - 1 marks then add the nth mark as follows. Place it first at the far right and let it move one position to the left in successive permutations until it reaches the far left. Then it stays there for one permutation before moving right one position at a time. When it reaches the right hand end it stays there for one permutation before starting to move back left again and so on.

(v) Fike's order. This is based on a one-to-one correspondence between the n! permutations of $1, 2, \ldots, n$ and the n! factorial representations $(d_2, d_3, \ldots, d_n)$ where d_k denotes an integer such that $0 \le d_k < k$. Starting with the original mark order 1 2 3 ... n the permutation is obtained from the sequence $(d_2 \, d_3 \, \ldots \, d_n)$ by interchanging the mark in position k with the mark in position $d_k + 1$ for $k = 2, 3, \ldots, n$. For example, if $n = 5$ the sequence 0 2 1 4 will generate the permutation 2 4 3 1 5 by the series of moves

12345 → 21345 → 21345 → 24315 → 24315

Fike's order starts with the factorial representation $(1, 2, \ldots, n-1)$ giving the permutation 1 2 3 ... n and continues with the sequences $(1, 2, \ldots, n-2, n-2), \ldots, (1, 2, \ldots, n-2, 0)$, $(1, 2, \ldots, n-3, n-1), \ldots,$ $(0, 0, \ldots, 0, 0)$. The final permutation is therefore n 1 2 ... n-1 . At first glance it would seem that n-1 interchanges are required for each permutation. However if we generate them in Fike's order in most cases only two interchanges are necessary.

(vi) Repeated objects - lexicographical order. The method of ordering is the same as in (i), the difference being that each object may appear any number of times. In general the first permutation is 1 1 ... 1 1 , and the last mark is increased until it is n ; this permutation is followed by 1 1 ... 2 1 etc.

5.2.2 Generation of All Permutations in Lexicographical Order

This section examines in detail one of the simplest, but not necessarily the most efficient, methods of generating all the n-permutations of n marks which are taken to be the integers $1, 2, 3, \ldots, n$. Each new permutation is generated from the previous one in lexicographical order.

The lexicographical ordering in table 5.2 suggests the rule for going from one permutation to the next. Denote the digits $k_1 k_2 k_3 k_4$ and consider the situation when $2\,4\,3\,1$ is to be transformed into $3\,1\,2\,4$. All the possible permutations with 2 in the k_1 position have been made and now it must be increased by the least amount. This can be done by interchanging k_1 and k_3 to obtain

$$3\ 4\ 2\ 1$$

However this is not the next lexicographical permutation; to get that we reverse the order of $k_2 k_3 k_4$ to obtain $3\,1\,2\,4$.

The generalisation of this example to the case of a permutation $k_1 k_2 \ldots k_n$ of n digits uses the following rules for obtaining the next permutation in lexicographical order:

Step 1. Find the largest i such that $k_{i-1} < k_i$.
Step 2. Find the largest j such that $k_{i-1} < k_j$.
Step 3. Interchange k_{i-1} and k_j.
Step 4. Reverse the order of the digits $k_i k_{i+1} \ldots k_n$.

For our example of transforming $2\,4\,3\,1$ we have $i = 2$, $j = 3$, thus interchanging k_1 and k_3 to get $3\,4\,2\,1$; then we reverse the digits $k_2 k_3 k_4$ to get $3\,1\,2\,4$.

A detailed algorithm for this method is given below.

```
procedure PERM LEX (integer array K(*); integer value N;
            logical value result L; procedure FINISH);
```
comment if on entry the array K contains a certain permutation of the N digits $1, 2, \ldots, N$ then the procedure will alter K so that it contains the next lexicographical permutation. If the first permutation $1\,2\,3 \ldots N$ is required, L should be set **true** on entry and the procedure will reset it

to false after it has put the first permutation in array K. If on entry K contains the last permutation $N(N-1)(N-2)\ldots 21$ then the procedure will exit using FINISH;

```
begin integer X ;
        if L then begin for i:=1 until N do K(i):=i;
                        L:= false; goto EXIT
                end;
for :=N step -1 until 2 do
if K(i) > K(i-1) then
begin X:=K(i-1);
        for j:=N step -1 until i do
        if K(j) > X then
        begin K(i-1):=K(j); K(j):=X;
                for m:=0 until (N-i-1) DIV 2 do
                begin X:=K(N-m); K(N-m):=K(i+m); K(i+m):=X
                end;
                goto EXIT
        end
end;
FINISH;
EXIT: end PERM_LEX;
```

5.2.3 The Adjacent Mark Method

In this method the permutation is formed from the previous one by interchanging mark k with its left or right hand neighbour according to whether $b_{k-1} + (k-1)b_{k-2}$ is even or odd, where $(b_n, b_{n-1}, \ldots, b_2)$ is the factorial representation of the mth permutation as defined by equation (5.4). A brief description of the adjacent mark method was given in section 5.2.1 together with the resulting ordering of the permutations in table 5.2. An algorithm for generating all the permutations by this method is given below.

```
procedure perm_adj_marks (integer array a, d, e(*);
                        integer value n; logical value result L);
begin integer q, v, x;
```

```
            if L then begin a(1):=1;
                            for i:=2 until n do
                            begin a(i):=i; d(i):=i; e(i):=-1
                            end;
                            L:=false; goto EXIT
                    end;
            v:=0;
            for k:=n step -1 until 2 do
            begin d(k):=q:=d(k) + e(k);
                    if q=k then e(k):=1
                        else if q=0 then begin e(k):=1; v:=v+1
                                        end
                        else begin q:=q+v; x:=a(q);
                                a(q):=a(q+1); a(q+1):=x;
                                goto EXIT
                            end
            end;
            L:=true;
EXIT: end perm_adj_marks;
```

This algorithm is considerably quicker than the one given previously
using lexicographical ordering, but it is more difficult to understand. The
problem basically is to find the two marks which have to be interchanged.
The array d , which is a representation similar to, but not the same as,
that given in equation (5. 4), indicates how far the mark has already moved
across the permutation. The array e is used to denote whether the mark
is being moved to the left (-1) or to the right (+1). Since mark n stays
at either end position for one permutation whilst the remaining marks are
permuted and similarly for mark n-1 etc. , we need on these occasions
to supplement the marks being interchanged. This is the purpose of the
variable v . In the arrays d and e , d_k and e_k refer to mark k
where $2 \leq k \leq n$; k is never equal to 1 since mark 1 is not a
travelling mark. In order to see the working of this algorithm more
clearly the values of d, e, q, v and a are given in table 5. 3 for n = 4 .
The logical variable L is used in a similar manner to the previous
algorithm: L is true at the start of the algorithm and the initial values

110

for the arrays a, d and e are set up. Subsequently L is false until the last permutation is generated when it is reset true.

d			e			v	a
d_2	d_3	d_4	e_2	e_3	e_4		permutation
2	3	4	-1	-1	-1	0	1 2 3 4
2	3	3	-1	-1	-1	0	1 2 4 3
2	3	2	-1	-1	-1	0	1 4 2 3
2	3	1	-1	-1	-1	0	4 1 2 3
2	2	0	-1	-1	1	1	4 1 3 2
2	2	1	-1	-1	1	0	1 4 3 2
2	2	2	-1	-1	1	0	1 3 4 2
2	2	3	-1	-1	1	0	1 3 2 4
2	1	4	-1	-1	-1	0	3 1 2 4
2	1	3	-1	-1	-1	0	3 1 4 2
2	1	2	-1	-1	-1	0	3 4 1 2
2	1	1	-1	-1	-1	0	4 3 1 2
1	0	0	-1	1	1	2	4 3 2 1
1	0	1	-1	1	1	0	3 4 2 1
1	0	2	-1	1	1	0	3 2 4 1
1	0	3	-1	1	1	0	3 2 1 4
1	1	4	-1	1	-1	0	2 3 1 4
1	1	3	-1	1	-1	0	2 3 4 1
1	1	2	-1	1	-1	0	2 4 3 1
1	1	1	-1	1	-1	0	4 2 3 1
1	2	0	-1	1	1	1	4 2 1 3
1	2	1	-1	1	1	0	2 4 1 3
1	2	2	-1	1	1	0	2 1 4 3
1	2	3	-1	1	1	0	2 1 3 4

Table 5.3   The values used in the adjacent mark permutation algorithm
for n = 4

5.2.4 The kth Permutation in Lexicographical Order

The previous two sections have shown methods of generating all the permutations in some order. A related problem is to find the kth permutation in some order given only the value k. For simplicity we consider the lexicographical order. Given n marks we can use the fact that the first (n-1)! permutations have the first mark in the first position, the next (n-1)! permutations have the second mark in the first position, and so on. Thus by integer division of k we find the mark which goes into the first position. This mark must be removed from the array of marks, the others moved up and the value of k reduced by an appropriate amount. When this is done, the operation above is repeated to find the mark in the second position. When n-1 positions have been filled there is only the last mark left. The algorithm for this method is shown below, it basically follows the description given above. Since the first permutation is marks(1), marks(2),..., marks(n), we divide (k-1) instead of k by q. It should be noted that q is (n-1)! first time round the for statement and subsequently becomes (n-2)!, (n-3)! etc.

```
procedure K_PERM (integer array a, marks (*); integer value k, n);
comment the array marks (1:n) contains the first mark in mark(1) etc.
The procedure finds the kth permutation in lexicographical order and the
resulting permutation is left in a(1:n) . The lexicographically first
permutation has a(i) = marks(i), i=1,...,n ;
begin integer q, c;
      q:=1; for i:=1 until n-1 do q:=q*i;
      for j:=1 until n-1 do
      begin c:=((k-1) div q) + 1;
            a(j):=marks(c);
            for i:=c until n-j do
            marks(i):=marks(i+1);
            k:=k-q*(c-1);
            if j=n-1 then a(n):=marks(1)
                     else q:=q div (n-j)
      end
end K_PERM;
```

Example. Consider $n = 4$ and marks(1:4) as $1, 2, 3, 4$. Then the 16th permutation is found by dividing 15 by 6 giving c=3, a(1)=3, marks(1:3) $= 1, 2, 4$. k is reduced to 4. Dividing 3 by 2 gives $c = 2$, $a(2) = 2$, marks(1:2) $= 1, 4$. k is reduced to 2. Dividing 1 by 1 gives $c = 2$, $a(3) = 4$, marks(1) $= 1$. Thus we obtain a permutation in a(1:n) of $3\,2\,4\,1$ in accordance with table 5.2.

5.2.5 Random Permutations

Since there are n! permutations of n marks, a random choice is one which gives each permutation probability $1/n!$ of being selected. Such random permutations are often required in practical applications, particularly in computer simulation or Monte Carlo experiments. Simulating scheduling strategies to decide in which order jobs should be performed is a typical example. Some experimental designs require random permutations of the treatments, sampling studies of certain distributions need random samples from the population of permutations, as does some work on non-parametric tests. There are many cases when n is so large that generating all the permutations is quite impractical and sampling the permutations is the only way to proceed.

The conceptually simplest method is to produce a random integer k in the range 1 to n! and then to generate the kth permutation in some convenient ordering. A standard pseudo-random number generator is likely to be used to obtain a random fraction, r (say), which it is hoped is uniformly distributed between 0 and 1; if r were so distributed, it can give a suitable random integer k in the required range 1 to n! by

$$k := \text{truncate } (n! \times r) + 1 \qquad\qquad (5.5)$$

Then an algorithm such as that given in section 5.2.4 can be used to find the kth permutation.

However, n! becomes large very quickly (e.g. greater than the largest integer that can be contained in a single word in storage) and this method has serious practical drawbacks since the source of the 'random' numbers may not even be able to produce every integer in $(1, n!)$ using equation (5.5), let alone to give them with equal frequency. Moreover the process of selecting a random k from the large range and then constructing the kth permutation in some order is unlikely to be any quicker in computation than an alternative method which makes several calls upon the random number generator, each to produce a small integer from which the random permutation is built. This alternative method is described below.

Assume we have $a_1, a_2, \ldots, a_n$ as the n objects in the boxes $1, 2, \ldots, n$; and we also have available (from a well-tested pseudo-random number generator, for example) a source of independent uniformly distributed variates in the range $(0, 1)$, $\tau_1, \tau_2, \tau_3, \ldots$ The method is:

Select a random integer i in the range $(1, n)$ by $i = [n\tau_1] + 1$ (where $[x]$ denotes the integer part of x), interchange the contents of boxes i and n .

Select a random integer j in the range $(1, n-1)$ by $j = [(n-1)\tau_2] + 1$ and interchange the contents of the boxes j and n-1 . (It is possible for $j = i$ so that a_n can be moved more than once.) Continue this process in the ranges $(1, n-2), (1, n-3), \ldots, (1, 2)$. A possible algorithm for this method is:

```
procedure random_perm (integer value n; integer array a(*);
                              real procedure random);
comment array a is 1:n and contains any permutations of the marks on
entry and a random permutation on exit.  The real procedure random when
called gives a random number in the range (0, 1) ;
begin integer temp, i;
    for k:=n step -1 until 2 do
    begin temp:=a(k); i:=truncate (k*random) + 1 ;
          a(k):=a(i); a(i):=temp
    end
end random_perm;
```

The method has close affinities with simple selection sort in which we start by swapping the item with largest key value (which is at a random place) with a(n) ; in the random permutation method we start by swapping a random value a(i) with a(n) . A moment's reflection will convince the reader that sorting and producing a random permutation are inverse operations - one starts with a random permutation and ends with a standard order, while the other ends with a random permutation but starts in order. The procedure can be used repeatedly to get several random permutations - there is no need to reinitialise array a because the method works equally well on an initial random permutation of the n marks. This method also uses interchanges very similar to those used in generating permutations in Fike's order.

5.2.6 Permutations with Repeated Marks

We consider now cases where the marks being permuted are not necessarily distinct. In the general case suppose the n objects consist of m_1 of the first type, m_2 of the second type and so on, and look at the simple case where each object can be repeated any number of times. As in section 5.2.2 the marks are assumed to be $1, 2, \ldots, n$ in that order. For $n = 4$ the final column of table 5.2 gives the first few permutations; if we study them carefully we can derive a general rule. The starting permutation is obviously $11 \ldots 11$; the last mark is increased by one until $11 \ldots 1n$ is reached. At this stage it is no longer possible to increase the last mark and one must be added to the (n-1)th mark, resetting the nth mark to 1 , giving $11 \ldots 21$. This idea is easily generalised to obtain an r-permutation of n objects with repetition from the previous one. Consider the r-permutation $k_1 k_2 \ldots k_r$; the algorithm to give the next permutation is:

1. Find the largest i such that $k_i < n$.
2. Add 1 to k_i .
3. Set $k_{i+1} = k_{i+2} = \ldots = k_r = 1$.

We note that if $i = r$, as it will in many cases, then the third step of the algorithm does nothing.

procedure PERM_REP (integer array K(*); integer value N, R;
                     logical value result L; procedure FINISH);
comment the parameters are the same as those used in generating permutations of distinct objects in lexicographical order, see section 5.2.2. The one additional parameter is R the size of the array K , so this procedure finds an R-permutation of the N objects $1, 2, \ldots, N$ with

repetition. If L is true it finds the first permutation $11 \ldots 1$ while subsequent entries with L false derive the next permutation in lexicographical order. If on entry K contains the last permutation $N N \ldots N$ then the procedure will exit using FINISH;

begin

    if L then begin for i:=1 until R do K(i):=1;

                L:=false; goto EXIT

            end;

    for i:=R step -1 until 1 do

    if $K(i) \neq N$ then begin K(i):=K(i)+1;

                      for j:=i+1 until R do

                      K(i):=1;

                      goto EXIT

              end;

    FINISH;

EXIT: end PERM_REP;

5.3 THE GENERATION OF COMBINATIONS

The problem of generating combinations does not arise in practice as frequently as that of generating permutations but sometimes before the generation of permutations we need to select the objects to be permuted from a larger population.

Consider generating r-combinations of the n objects $1, 2, \ldots, n$ in lexicographical order. For example in selecting four digits from the seven digits $1, 2, \ldots, 7$ the first twelve combinations are

1. 1234	5. 1245	9. 1257
2. 1235	6. 1246	10. 1267
3. 1236	7. 1247	11. 1345
4. 1237	8. 1256	12. 1346

Let a general combination be $k_1 k_2 k_3 k_4$; since they are in lexicographical order $k_4 \leq 7$, $k_3 \leq 6$ etc. In combination 10 note that k_4 and k_3 cannot be increased to derive combination 11 so k_2 is increased by 1 , while k_3 is made one greater than k_2 , and k_4 one greater than k_3. Generalising this for finding the next r-combination from the previous one $k_1 k_2 \ldots k_r$ suggests the algorithm

1. Find the largest i such that $k_i < n-r+i$.
2. Add 1 to k_i.
3. Perform the substitutions $k_j = k_{j-1} + 1$ for $j = i+1, i+2, \ldots, r$.

The detailed program for this is given below:

```
procedure COMB_LEX (integer array K(*); integer value N,R;
                    logical value result L; procedure FINISH);
comment if on entry the array K(1:R) contains a combination of the N
digits 1,2,...,N then the procedure will alter K so that it contains the
next lexicographical combination. If the first combination 1 2 ... R is
required L should be true on entry, L will be reset false and should
remain so for all subsequent entries. If K contains the last combination
N-R+1 N-R+2 ... N then the procedure will exit using FINISH;
begin
        if L then begin for i:=1 until R do K(i):=i;
                        L:=false; goto EXIT
               end;
        for i:=R step -1 until 1 do
        if K(i) < N-R+i then begin K(i):=K(i)+1;
                                   for j:=i+1 until R do
                                   K(j):=K(j-1)+1;
                                   goto EXIT
                            end;
                FINISH;
EXIT: end COMB_LEX;
```

The combinations with repeated marks can be generated in a similar manner to the method used for permutations in section 5.2.6. (See exercise [5.8].)

5.4   THE GENERATION OF COMPOSITIONS

Section 4.5 defined a composition as an ordered division of a positive integer n into parts $a_1, a_2, \ldots, a_m$. The number of parts in an unrestricted composition can vary. In section 4.5.1 there is a method of enumerating these unrestricted compositions which can also be used as the basis of an algorithm for generating them. There is a one-to-one correspondence between the compositions and the integers 0 to $2^{n-1}-1$. An integer in this range can be expressed as an $(n-1)$ bit binary number, and its zeros and ones can be placed in the $(n-1)$ spaces in a row of n marks, X's say. The ones of the binary number show the divisions into the parts of the corresponding composition.

For example, let $n = 7$ and consider the integer 13 as a 6-bit binary number; $(13)_{10} = (001101)_2$. The six spaces between the seven X's are divided as shown:

$$(X_0 X_0 X)_1 (X)_1 (X_0 X)_1 (X)$$

which gives the composition $(3\,1\,2\,1)$.

The number $n = 0$ therefore corresponds to the composition (n) which has just one part, whilst the number $2^{n-1} - 1$ corresponds to the composition $1\,1\,1 \ldots 1$ with n parts each of size 1. The method of representing the composition $(3\,1\,2\,1)$ of 7 by the binary pattern 001101 is known as the difference representation.

The algorithm given below for unrestricted compositions takes each integer in turn from 0 to $2^{n-1} - 1$ and finds its binary equivalent by repeated division by 2; as it generates the appropriate binary digits it forms the parts of the composition - each time integer division by 2 leaves a remainder of 1 the next part can be found (see the else part of the procedure below).

procedure COMPOSITIONS (integer value n);
comment this procedure generates all the $2**(n-1)$ compositions of a given integer n. A one-to-one correspondence is established between the compositions and the integers I from 0 to $2**(n-1)-1$. The I integers are considered as $(n-1)$ bit binary numbers with the ones forming the divisions. The composition is set up in the array c ;

```
begin integer j, y, z, s, sum; integer array c (1::n);
        for I:= 0 until ROUND(2**(n-1)-1) do
        begin y:= I; sum:= 0; s:=j:=1;
                while y> 0 do
                begin z:= y REM 2; y:= y DIV 2;
                        if z=0 then s:=s+1
                        else begin c(j):=s; j:=j+1;
                                        sum:=sum+s; s:=1
                                end
                end;
                c(j):=n - sum;
                comment the composition is now available in the array c from
                c(j) to c(1);
        end
end COMPOSITIONS;
```

The generation of the compositions of n with exactly m parts is
not quite so straightforward as that for unrestricted compositions. We can
of course adapt the above algorithm so that it rejects all the $(n-1)$ binary
numbers except those with exactly $m-1$ ones but this would be wasteful.
Another alternative would be to generate $\binom{n-1}{m-1}$ combinations, and use
them to select the $(m-1)$ positions out of a possible $(n-1)$ where we put
the ones.

The algorithm given below to find all the compositions of n with
exactly m parts is recursive. It sets $A(m)$ to each of its possible
values $1, 2, \ldots, n-m+1$ in turn and, for each such value, generates
(recursively) all the compositions of $n-A(m)$ with $m-1$ parts. The
ordering obtained for the compositions is reverse lexicographical, so if
true lexicographical ordering is required the composition can be read
$A(m) A(m-1) \ldots A(2) A(1)$.

```
procedure COMP (integer value n, m; integer array A(*); procedure PROC);
comment this procedure generates all the compositions of n with exactly
m parts. The compositions are generated in reverse lexicographical
order;
if m≤1 then begin A(1):=n; PROC;
```

> comment we can now use the composition which is in
> A(1) ... A(m) by means of the procedure PROC;
> end
else for i:=1 until n-m+1 do
> begin A(m):=i; COMP (n-i, m-1, A, PROC)
> end COMP;

5.5 THE GENERATION OF PARTITIONS

Partitions are the unordered divisions of an integer into parts, i. e. 5 2 1 1 is the same partition of 9 as 1 2 5 1. We will therefore generate the partitions of n into an array $c(1), c(2), \ldots, c(n)$ such that $c(1) \geq c(2) \geq \ldots \geq c(n)$. The first partition is (n) itself, the next (n-1 1) and the last one is (1 1 ... 1). Given any partition the next one is formed by subtracting 1 from the rightmost part c(k) which is greater than 1 and then the remainder $(n - \sum_{j=1}^{k} c(j))$ distributed as quickly as possible.

Thus if the remainder is less than or equal to c(k) it is all put into c(k+1); alternatively c(k+1) is made the same as c(k) and the new remainder distributed as quickly as possible. For example consider the partition 5 4 1 1 1 : the rightmost part greater than 1 is 4, so this part is reduced by 1 giving 5 3 with a remainder of 4. When this remainder is distributed as quickly as possible we obtain the final partition 5 3 3 1, since the most that can be put in the third part is 3 (the third part is less than or equal to the second part) and this leaves a new remainder of 1 which can be put as the fourth part. The detailed algorithm for this method is given below.

procedure PARTITIONS (integer value n);
comment this procedure derives all the partitions of the integer n starting with n itself and finishing with 1 1 ... 1. The partition is in $c(1), c(2), \ldots, c(parts)$ with $c(i) \geq c(j)$ if $i < j$. A partition is derived from the previous one by subtracting 1 from the rightmost part greater than 1. The remainder, rdr, is distributed in c(k+1), c(k+2) etc. as quickly as possible;
begin integer parts, rdr, q; integer array c(1::n);
> c(1):=n; parts:=1;
> comment this is the first partition n with one part;

120

```
RESTART: for k:= parts step -1 until 1 do
           if c(k) ≠ 1 then
           begin q:=c(k):=c(k)-1; rdr:=parts-k+1;
                 for j:=k+1 until n do
                 if q < rdr then begin c(j):=q; rdr:=rdr-q
                                 end
                 else begin c(j):=rdr; parts:=j;
                             comment the next partition is in
                             c(1), c(2),..., c(parts)  and may be
                             used;  goto RESTART;
                      end
           end;
     end PARTITIONS;
```

There are many other algorithms for generating all the partitions of an integer, and an example of a recursive version is given below.

```
procedure RECURSIVE_PART (integer value remd, p;
                          integer array a(*));
if remd=0 then comment use the partition in a(1), a(2),..., a(p-1)
          else begin integer k;
               k:=if p=1 or remd ≤ a(p-1) then remd
                                          else a(p-1);
               for j:=k step -1 until 1 do
               begin a(p):=j; RECURSIVE_PART (remd -j, p+1, a);
               end
          end RECURSIVE_PART;
```

Notice that remd is what is left over at each stage after we have assigned $a(1), a(2), \ldots, a(p)$; if it is zero then the partition is ready to use. The procedure is called initially with remd and p equal to n and 1 respectively. The partitions of n are then obtained in reverse lexico-graphical order starting with n , then (n-1 1) , and finishing with 1 1 ... 1 .

5.6 RECURRENCE RELATIONS

Recurrence relations, or difference equations, which have been established to calculate the number of configurations of a particular type can often also be used to generate those configurations and to answer some questions about the order so derived. For example, one proof of the elementary result that the number of permutations of n distinct objects (here taken as the integers $1, 2, \ldots, n$) is $u_n = n!$ notices that $u_1 = 1$ and $u_n = nu_{n-1}$, since each permutation of the first n-1 integers can yield n permutations of $1, 2, \ldots, n$ by inserting the integer n into any of the n-2 spaces between the integers or at the beginning or end of the (n-1)-permutation. Thus, the method of derivation of the difference equation indicates recursive methods of listing the permutations; the order obtained by inserting the new item from the rightmost positions first are:

n = 2	n = 3	n = 4
1 2	1 2 3	1 2 3 4
	1 3 2	1 2 4 3
	3 1 2	1 4 2 3
		4 1 2 3
		. . . .
		. . . .
2 1	2 1 3	2 3 1 4
	2 3 1	2 3 4 1
	3 2 1	2 4 3 1
		4 2 3 1
		. . . .
		. . . .

Identification of the position of a given n-permutation in this ordering is itself obtained recursively; if the number of places from the right which the integer n occupies is r_n, the permutation is at position

$$v_n = r_n + n(v_{n-1} - 1) \quad \text{and} \quad v_1 = 1$$

where v_{n-1} is the position of the (n-1)-permutation obtained by deleting n in the ordering of all (n-1)-permutations. For example, the sequence of calculations necessary to find the position of 2 1 4 3 in the order is:

$r_4 = 2$ since 4 is in the second position from the right in 2 1 4 3

$r_3 = 1$ since 3 is in the first position from the right in 2 1 3

$r_2 = 2$ since 2 is in the second position from the right in 2 1

$v_2 = 2 + 2(v_1-1) = 2$

$v_3 = 1 + 3(2-1) = 4$

$v_4 = 2 + 4(4-1) = 14$

The converse problem to derive the permutation for a given position is solved by finding successively the positions of $n, n-1, n-2, \ldots$ For example, for $n=4$ we get the fourteenth permutation by noting $14/4 = 3 +$ remainder 2 . Hence 4 is in the second position from the right .. 4. ; and $(v_3-1) = 3$ so that the 3-permutation is fourth in its order; then $4/3 = 1 +$ remainder 1 and so we have .. 3 for the position of the 3 , giving .. 43 and finally $(v_2-1) = 1$ which yields 2 1 4 3 as the fourteenth in the order. A remainder of zero at any stage indicates that the digit being considered is at the extreme left position and the dividend should be reduced by 1 .

The ordering produced in this way is thus the same as that based upon the representation of integers in the form

$$N \equiv n! \left[\frac{a_n}{n!} + \frac{a_{n-1}}{(n-1)!} + \cdots + \frac{a_2}{2!} \right]$$

with $0 \le a_i < i$, where a one-to-one correspondence between the integers $\{0, 1, \ldots (n!-1)\}$ and all n-permutations is obtained by regarding a_i as the number of digits less than i which appear to the right of it. Thus, the fourteenth permutation follows from $13 \equiv (a_2, a_3, a_4) = (1, 0, 1)$.

More complex problems which lead to second (and higher) order equations can also be treated.

Consider the number P_k of permutations of k objects drawn with whatever repetition is desired from n distinct objects $(1, 2, 3, \ldots)$ such that no three adjacent objects in the permutations shall be the same (cf. exercise [2.13]).

These k-permutations can be divided into two sets according as their last two objects are the same or are different. Any k-permutation with the last two objects the same can be obtained from a (k-2)-permutation by attaching two like symbols to it, as long as they are distinct from the previous final object; there are thus $(n-1)P_{k-2}$ such k-permutations.

123

Similarly all k-permutations of the other set are $(n-1)P_{k-1}$ in number and are obtained by attaching one different symbol to the end. Hence

$$P_k = (n-1)(P_{k-2} + P_{k-1})$$

where clearly $P_1 = n$, $P_2 = n^2$ and an expression for P_k follows.

Once again, a systematic ordering is derived naturally, but this time the ordering for k-permutations involves both (k-1)- and (k-2)-permutations and to be able to start an ordering initial orders of the permutations of one element and of two elements are required. For example, for $k = 3$, appropriate orders to start with $n = 1, 2$ are shown under $P_1 = 3$, $P_2 = 9$; the first 6 of the $P_3 = 24$ permutations for $n = 3$ are derived from the P_1 column and the remaining 18 correspond in pairs to those in the P_2 column.

Listing of restricted permutations

$P_1 = 3$	$P_2 = 9$	$P_3 = 24$			
1	1 1	1 2 2	1 1 2	2 2 3	
2	1 2	1 3 3	1 1 3	2 3 1	
3	1 3	2 1 1	1 2 1	2 3 2	
	2 1	2 3 3	1 2 3	3 1 2	
	2 2	3 1 1	1 3 1	3 1 3	
	2 3	3 2 2	1 3 2	3 2 1	
	3 1		2 1 2	3 2 3	
	3 2		2 1 3	3 3 1	
	3 3		2 2 1	3 3 2	

The permutations for $k = 3$ can then be listed by attaching the possible identical pairs of objects in turn to the P_1 members, followed by those (P_2) with the possible single object added, and similarly for $k > 3$.

The identification of the place in the order occupied by a given one of these restricted k-permutations is attained by determining successively which permutation it is derived from, i.e. whether it comes from a (k-1)- or a (k-2)-permutation and so on until the permutations of one or two elements are reached. Figure 5.2 indicates the process for a 5-permutation where D, D' denote different digits.

124

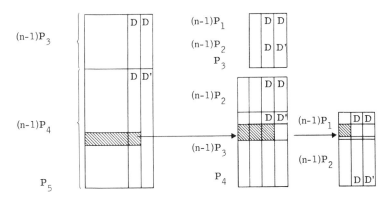

Fig. 5.2. Identification of a restricted permutation

The reverse process of finding the permutation at a given place in the order consists of repeated division by $(n-1)$, subtraction of P_{k-2} if the quotient exceeds it and noting the remainders to permit the building of the permutation from whichever member of the P_1 1-permutations or P_2 2-permutations is appropriate.

The above examples should have alerted the reader to the possibilities of deriving methods of generating configurations whenever suitable recurrence relations are available.

5.7 BIBLIOGRAPHY

The two survey papers by Lehmer [1, 2], although now both well over ten years old, still contain much relevant material and are very readable. The other good general reference is Wells [3] with the reservation mentioned previously that it is not an easy book to read.

The efficient generation of permutations is such an important problem in computing that it has received considerable attention in the literature. A recent survey article by Sedgewick [4] contains a good bibliography of the subject. The algorithm given here for finding all the permutations in lexicographical order was adapted from that of Shen [5]; the adjacent mark method was discovered independently by Johnson [6] and Trotter [7], and generation by transposition is in Wells [8]. Fike [9] gave the ordering with which his name is associated and a corresponding algorithm which has since been improved by Rohl [10]. The algorithm for generating random permutations is due to Page [11].

The generation of combinations seems to arise relatively infrequently in computer applications and less attention has been directed to the construction of efficient algorithms. The one given here for generating combinations in lexicographical order is also adapted from Shen [5].

There are several methods of generating compositions and partitions. The one given for compositions is the same as that explained by Lehmer [2]. Recursive techniques can also be used for generating compositions as well as for partitions.

[1]   D. H. Lehmer: 'Teaching Combinatorial Tricks to a Computer.' pp179-93 in 'Proceedings of Symposia in Applied Mathematics, 10. Combinatorial Analysis' edited by R. E. Bellman and Marshall Hall, Jr. American Math. Soc., Providence, R. I., 1960.

[2]   D. H. Lehmer: 'The Machine Tools of Combinatorics.' Chapter 1, pp5-31, in 'Applied Combinatorial Mathematics' edited by E. F. Beckenbach. John Wiley, 1964.

[3]   M. B. Wells: 'Elements of Combinatorial Computing.' Chapter 5, pp127-57. Pergamon Press, 1971.

[4]   R. Sedgewick: 'Permutation Generation Methods.' ACM Computing Surveys, Vol 9, No 2 (June 1977), pp137-64.

[5]   Mok-kong Shen: 'On the Generation of Permutations and Combinations.' BIT, Vol 2 (1962), pp228-31.

[6]   S. M. Johnson: 'Generation of Permutations by Adjacent Transposition.' Math. Comp., Vol 17 (1963), pp282-5.

[7]   H. F. Trotter: 'ACM Algorithm 115 - Perm.' Comm. ACM, Vol 5 (1962), pp434-5.

[8]   M. B. Wells: 'Generation of Permutations by Transposition.' Math. Comp., Vol 15 (1961), pp192-5.

[9]   C. T. Fike: 'A Permutation Generation Method.' Computer J., Vol 18, No 1 (Feb 1975), pp21-2.

[10]   J. S. Rohl: 'Programming Improvements to Fike's Algorithm for Generating Permutations.' Computer J., Vol 19, No 2 (May 1976), pp156-9.

[11]   E. S. Page: 'A Note on Generating Random Permutations.' Applied Statistics, Vol 16, No 3 (1967), pp273-4.

5.8 **EXERCISES**

[5.1] The integers m $(0 \le m \le 5039)$ are represented by the factorial digits $(a_1, \ldots, a_6)$ where

$$m = a_1 1! + a_2 2! + \ldots + a_6 6!$$

and

$$0 \le a_i \le i$$

(a) What are the decimal values of m for the representations $(0, 1, 2, 3, 4, 5)$, $(0, 1, 0, 1, 0, 1)$ and $(1, 2, 3, 3, 2, 1)$?

(b) What are the factorial representations $(a_1, a_2, \ldots, a_6)$ for the integers 3000, 100 and 1235 ?

[5.2] Given $n = 5$ and the five marks 0, 1, 2, 3, 4 what are the 50th and 100th permutations in the following orders:

(a) Lexicographical;

(b) Reverse lexicographical;

(c) Adjacent mark order;

(d) Fike's order?

(In each case the first permutation is 0 1 2 3 4 .)

[5.3] Write a procedure similar to that given in section 5.2.2 which will generate all the permutations of n in reverse lexicographical order.

[5.4] Fike's method, which was explained briefly in section 5.2.1, uses the one-to-one correspondence between the n! permutations and the n! sequences $(d_2, d_3, \ldots, d_n)$ of the factorial representation where $0 \le d_k < k$.

Fike's method derives the permutations from the sequences $(d_2, d_3, \ldots, d_n)$ as follows:

(i) Start with the permutation 1 2 3 ... n and the sequence $(1, 2, \ldots, n-1)$.

(ii) Find the next lower lexicographical sequence; the next permutation is obtained from the sequence by interchanging the marks in positions k and $d_k + 1$, $k = 2, 3, \ldots, n$.

The first few sequences and permutations for $n = 5$ are

Sequence	Permutation
1, 2, 3, 4	1 2 3 4 5
1, 2, 3, 3	1 2 3 5 4
1, 2, 3, 2	1 2 5 4 3
1, 2, 3, 1	1 5 3 4 2
1, 2, 3, 0	5 2 3 4 1
1, 2, 2, 4	1 2 4 3 5
.	.
.	.
.	.

The final sequence is $(0, 0, 0, 0)$ which gives the permutation 5 1 2 3 4 .
Write a procedure to generate all the permutations in Fike's order.

[5.5] Write in lexicographical order the r-permutations of n objects when $r = 3$ and the objects are 1, 2, 3 and 4 . Give an algorithm for generating such permutations for general values of r and n .

[5.6] The r-combinations of n objects with unlimited repetition can be represented by a binary sequence with n+r-1 bits containing r ones, see for example section 4.2.1, equation (4.10). Write a procedure to generate the binary sequences and from them the r-combinations with repetition. For $r = 3$ and the four objects 1, 2, 3, 4 the first few sequences and combinations are:

Binary sequence	Combination
(000111)	1 1 1
(001011)	1 1 2
(010011)	1 1 3
(100011)	1 1 4
(001101)	1 2 2
(010101)	1 2 3
(100101)	1 2 4
(011001)	1 3 3
.	.
.	.
.	.

[5. 7] Write a procedure which, given n distinct marks (e. g. 0, 1, 2, 3, ..., n-1) and a number m $(0 \le m < n!)$, will derive the mth lexicographical permutation of the n marks. Explain how you would test your procedure. The zeroth permutation is 0 1 2 3 4 ... n-1. [Hint: One of the better methods is to express m in the form

$$m = a_1 1! + a_2 2! + ... + a_{n-1} (n-1)!$$

where $0 \le a_i \le i$ and use Lehmer's lexicographical method.]

(Newcastle 1967)

[5. 8] Write down what you understand by the lexicographical ordering of combinations of n distinct objects taken r at a time with unrestricted repetition.

Write a procedure to generate these combinations. The procedure should be such that on initial entry it generates the first combination and on each subsequent entry it finds the next lexicographical combination from the previous one. If entry is made with the last combination it should exit.

(Newcastle 1969)

[5. 9] Write a program which generates random compositions of n.

[5. 10] Write procedures which generate the following restricted partitions:

   (a) The partitions of n into exactly k parts.

   (b) The binary partitions of n. (A binary partition is one which contains only parts of size $1, 2, 4, ..., 2^i$.)

   (c) The perfect partitions of n. (A perfect partition of an integer n contains a partition of each number less than n in one and only one way; e. g. the perfect partitions of 7 are 4 2 1, 2 2 2 1, 1^7.)

[5. 11] Write a procedure which will enumerate the number of partitions of n. [Hint: It is suggested that the following recurrence relation for the unrestricted partitions of n be used:

$$p_n = p_{n-1} + p_{n-2} - p_{n-5} - p_{n-7} + \ldots + (-1)^{k+1}(p_{n-k_1} + p_{n-k_2}) + \ldots$$

where

$$k_1 = \frac{3k^2 - k}{2} \ , \ k_2 = \frac{3k^2 + k}{2}$$

Candidates are advised that it is very difficult to write an efficient recursive routine for this problem.]

(Newcastle 1971)

[5.12] Find the number of restricted r-permutations of n objects with unlimited repetition but with no adjacent elements the same.

6 · Search Procedures

BACKTRACK PROGRAMMING

6.1.1 Enumeration of Vectors

Backtrack programming is a method for the systematic enumeration of a set of vectors. It is applicable to discrete problems whose possible solutions can be described by vectors which need not necessarily all have the same dimensions but whose elements are members of certain finite sets. There could be great complexity involved in the definition of the sets from which the elements are drawn and considerable dependence of these definitions on other elements in the vector. Indeed in some respects the method is most effective when the complexity of such definitions can be exploited to reduce the size of the sets of elements which have to be considered at any stage. All the problems which have been encountered so far in this book fall into this category and the following examples give some illustrations of the vectors concerned.

Example 1. Permutations without repetition of n marks $1, 2, \ldots, n$. The enumeration of these permutations needs the enumeration of the vectors of n elements $(x_1, x_2, \ldots, x_n)$ where the first element is selected from the whole n marks, the next element from the n marks omitting the first element that has been chosen and so on.

$$x_1 \in \{1, 2, \ldots, n\}$$
$$x_2 \in \{1, 2, \ldots, n\} - \{x_1\}$$
$$x_3 \in \{1, 2, \ldots, n\} - \{x_1, x_2\}$$

Example 2. Combinations of r elements selected from n marks $1, 2, \ldots, n$. Each vector has r elements and, as the order is immaterial, we can suppose them arranged in lexicographical order. The first element can be any one of the n marks, the second element any one of the marks greater than the first element that has been chosen and so on.

$$x_1 \in \{1, \dots, n\}$$
$$x_2 \in \{x_1 + 1, x_1 + 2, \dots, n\}$$
$$x_3 \in \{x_2 + 1, x_2 + 2, \dots, n\}$$

Example 3. Compositions of an integer n. In this example the vectors do not all have the same number of elements but each is a vector of positive integers and there can be at most n elements. The first element can be any one of the first n integers while the second is selected from the first $n-x_1$ integers and so on.

$$x_1 \in \{1, \dots, n\}$$
$$x_2 \in \{1, 2, \dots, n-x_1\}$$
$$x_3 \in \{1, 2, \dots, n-x_1-x_2\}$$

Example 4. Partitions of n. As in example 3 the vectors can have variable numbers of elements and if the standard order is reverse lexicographical, the first element is chosen from the whole set of n integers but the second element only from the first x_1 as long as the sum of the first two elements does not exceed n. The sets are therefore defined as follows

$$x_1 \in \{1, 2, \dots, n\}$$
$$x_2 \in \{1, 2, \dots, \min(x_1, n-x_1)\}$$
$$x_3 \in \{1, 2, \dots, \min(x_2, n-x_1-x_2)\}$$

All problems such as the ones that we are considering can be represented graphically as a tree, each level corresponding to the insertion of another component of the vector. Thus in figure 6.1, the number (4) of branches leading to the first level and, of course, the number of nodes at that level, is the number of possible elements that can appear as the first component of the vector. At the second level, there are nodes corresponding to each of the possible elements in the

132

second component given the choice of the first component, for all possible first components (3, 2, 1, 0 in figure 6.1). In a general problem there could be different numbers of branches emerging from each node as well as different numbers of levels before the terminal nodes (the 'leaves') are reached which correspond to the possible complete vectors. In this example the vectors are (a_1, a_{11}, a_{111}), (a_1, a_{12}), (a_1, a_{13}), (b_1, b_{11}), (b_1, b_{12}, b_{121}), (c_1, c_{11}), (d_1).

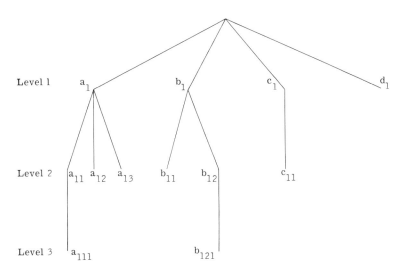

Level 1 a_1 b_1 c_1 d_1

Level 2 a_{11} a_{12} a_{13} b_{11} b_{12} c_{11}

Level 3 a_{111} b_{121}

Fig. 6.1. A tree of possible vectors

The backtrack method for generating the vectors systematically is best described informally by discussing its operation in the midst of the process. Suppose a complete vector $(x_1, x_2, \ldots, x_r)$ has just been constructed and perhaps made available to some further routine for processing. On return to the backtrack procedure an attempt is made to find a new rth element, from the set X_r, say, of elements which could occur in this position given the values of the elements that are in the first $r-1$ positions of the vector. If X_r is not empty its first member x_r' may be taken, deleted from the set and inserted in place of x_r. In this case we may have another complete vector or we may have to select further elements in the vector but the set X_r has been reduced by one member. If however X_r was empty it is necessary to 'backtrack'

to the next component of the vector and replace x_{r-1} ; this can be done
if the set of remaining possible members for that element X_{r-1} itself
is non-empty. If it has a member we can choose a new x_{r-1} , delete
that member from the set X_{r-1} and then form if necessary a new set
X_r of elements which are now possible contenders for the rth component
of the vector. If X_{r-1} was empty it would be necessary to backtrack
further.

6.1.2 Backtrack Algorithms

The basic algorithm for backtrack follows the description given in
the previous section. The vector is built up one element at a time;
suppose that a partial vector $(x_1, x_2, \ldots, x_{k-1})$ has been constructed;
then we construct the set X_k which includes all the candidates for the
next element x_k . There are now two possibilities:

(a) X_k is not empty. In this case we select the smallest element
in X_k (assuming therefore the elements of X_k are linearly ordered),
and then proceed in a similar manner to construct x_{k+1} .

(b) X_k is empty. We backtrack to reconsider X_{k-1} changing
x_{k-1} to the next element in X_{k-1} if one exists; if not we backtrack
further.

The logical structure of the flow diagram shown in figure 6.2 can be converted
into an Algol-like algorithm as follows:

$$k := 1; \; X_k := \{X_1\};$$

LOOP: while X_k is not empty do

    begin $x_k :=$ smallest element of the set X_k;

    if vector complete then comment use the vector

                      $(x_1, x_2, \ldots, x_k)$ just found and

                      if further vectors are required

                      remove x_k from the set X_k and

                      continue

              else begin k := k+1; determine the

                set X_k;

              end;

    end;

    if k = 1 then STOP

else begin comment this is where we backtrack;

k:=k-1; determine the set X_k;

remove from X_k all elements

$\leq x_k$; goto LOOP

end;

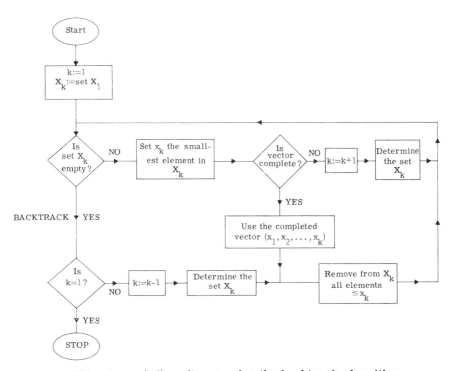

Fig. 6.2. A flow diagram for the backtrack algorithm

There are several variations of this basic algorithm, which itself leaves unanswered several important questions such as the most efficient way to determine X_k. Ideally X_k should be quickly set up and contain as few elements as possible. When we check to see if the vector is complete in many cases (such as r-permutations) this will be a simple test to see if $k=r$. However there are other cases (e.g. partitions) when the number of elements in the vector is variable and at any stage depends upon the elements so far constructed. It is obviously efficient if we can determine at an early stage in the construction of the partial vector that it is not suitable and so we rule out large areas of the search space without further examination. The general framework of the backtrack algorithm given above leaves the programmer considerable scope to exercise his skill in an individual problem, and because backtrack is an exhaustive search it is usually very necessary to try to minimise the number of operations.

Since backtrack programming is closely related to tree searching we can consider using recursive techniques in our implementations. Consider for example cases where we are examining the points p of a graph, and we have available an adjacency structure $A(p)$ of the points directly attached to p, then a recursive backtrack search could be expressed as follows:

```
procedure Backtrack (integer value p);
begin add point p to the stack S;
        while q ϵ A(p) do
        if q is not in the stack S then Backtrack(q);
        delete p from the stack S;
end;
```

Typically the shortest path algorithms discussed in section 4.6.4 could be expressed in this manner.

6.1.3 Gains and Losses

In spite of the simplicity of the idea of backtrack programming it is clear that in some problems the programs can be unavoidably long or complex and numerous calculations are required for the production of each acceptable vector. The complications which can arise in backtrack programs are principally in the formation of the sets of possible elements for the different components of the vectors - during these operations tests are being performed to see how the elements of the vector already chosen restrict the choice of elements for the next component. Thus the backtrack method aims to do all its testing for validity according to the specification of the problem during the formation of the vectors; it may be contrasted with the enumerative approach at the other extreme which generates all the complete vectors and only tests the complete vectors for validity. There are, of course, intermediate backtrack strategies which do some of the testing while forming the vectors and leave the rest until a complete one is obtained. Intuitively it seems that backtrack will gain when the reduction in the number of vectors produced is large and so outweighs the extra work during production of the vectors. This can be illustrated in one or two simple problems in enumerating permutations. Suppose, even having progressed so far through this book, that one were so naive as to plan to generate the permutations of the first n integers, $\{1, 2, \ldots, n\}$, by (i) forming all the n^n vectors of n elements, each element chosen from $\{1, 2, \ldots, n\}$ and (ii) then rejecting each vector in which there were any repeated elements. Thus if $n = 3$, the 27 vectors might be generated in lexicographical order starting $(1, 1, 1), (1, 1, 2)$, $(1, 1, 3), (1, 2, 1), \ldots$ and ending $(3, 3, 2), (3, 3, 3)$. Each vector would be tested and all but $3! = 6$ of them rejected. Even in this little example, $n = 3$, the naive procedure generates more than four times as many complete vectors as there are valid permutations, and the factor of four or so increases rapidly as the size of problem increases (for $n = 10$ the factor is more than two thousand, and it increases exponentially with n).

Clearly such a degree of naivety is expensive, as one might have expected, and backtrack programming can show gains even if its intermediate testing procedures need some effort. But combinatorial problems can grow so quickly that just a little more sophistication may not be enough. Consider, for example, the task of generating all permutations of $\{1, 2, \ldots, n\}$ which are such that each element i is no more than one place removed from the ith position - thus element i must be in the (i-1)th, ith or (i+1)th positions. One way - the 'nearly' naive way - would be to generate the $n!$ permutations and reject all those which broke the specifications. But in this problem too, the ratio of acceptable vectors (permutations in this case) to rejected ones is much too low for large problems. Indeed the number of acceptable permutations increases by a factor of about 1.6 as n increases by 1 while the total number of permutations to be tested increases by a factor n (cf. exercise [2.16]). It is quite feasible to generate by an efficient method all the permutations with the conditions stated for $n = 12$ but daunting to start by forming the 400 million or so unrestricted permutations and then to reject nearly all of them.

In a pictorial representation using a tree in which each level corresponds to an element in the vector, the backtrack procedure involves the pruning (or, perhaps more accurately, the stunting of the growth) of the tree. Thus in figure 6.3 the 27 free ends, or leaves, of the tree correspond to the $3^3 = 27$ vectors of the naive approach which starts to form all the permutations of 3 items with repetition. The single cut (-) on a branch shows those parts of the tree that would not be generated by the backtrack program for the unrepeated and unrestricted permutations, and the double cut (=) the further parts of the tree avoided by the backtrack program for the restricted permutations.

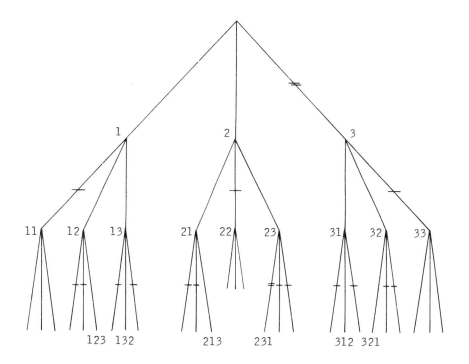

Fig. 6.3. The tree of permutations

In combinatorial problems where complete or partial enumeration is contemplated the following warning cannot be stressed too strongly: assess as best you can whether the magnitude of the task is within the range of the resources available.

6.2   BRANCH AND BOUND METHODS

6.2.1   Searching for a Optimum

Sometimes a complete or partial enumeration of vectors is undertaken because a list of those vectors is required for some purpose - for example, a list of <u>all</u> the solutions of a certain combinatorial problem. On other occasions the list may just be an intermediate stage in the work and subseq·ently the list will be searched to find one or more of the vectors that are optimum in a given sense. The idea is that of a 'cost' associated with each vector and one approach (again a naive one) to find the best vector or vectors according to the cost criterion is to enumerate them all, evaluate all their costs and pick the best ones. That task is clearly a bigger one than just doing the enumeration (which itself might be too big to be feasible, even by an efficient method). In some problems one might hope that a large proportion of the vectors would 'obviously' stand no chance of being an optimum and need not have their costs evaluated, and perhaps not even need be generated. We are led to take one of the ideas of backtrack programming a stage further - just as this method aims to avoid generating complete vectors which cannot satisfy the conditions of the problem, we want to avoid generating vectors for which the costs cannot possibly be optimum. To be able to do this, it must be possible to calculate a useful bound to the cost of a complete vector, $C(x_1, x_2, \ldots, x_n)$ say, when only part of the vector $(x_1, x_2, \ldots, x_r)$, $r < n$, has been formed. If there is a bounding function $B(x_1, x_2, \ldots, x_r)$ that can be calculated just from a knowledge of the first r elements of the vector which guarantees that the cost of a complete vector with the same first r elements is greater than $B(x_1, \ldots, x_r)$ then there is no point in generating any of those complete vectors if we are searching for one with least cost <u>and we already have found a complete vector with cost less than</u> $B(x_1, \ldots, x_r)$.

Such is the 'bounding' part of the 'branch and bound' approach to finding an optimum (or optima); the bounding prunes the tree of vectors to be generated and costed still further. In problems of any size the pruned trees are still substantial with many nodes and there will be many alternative sequences in which the nodes can be considered in order to generate the next level of nodes, i.e. the order in which the branches are constructed. In the next section we look at some alternative strategies and their implications for the computing resources needed.

6.2.2 Branching Strategies

Suppose that a branch and bound method is to be used for finding a vector that produces a minimum of a certain cost function and that it has been possible to find a bound function which can be calculated for every node of the tree (except, of course, the leaves where the cost function itself can be calculated) and which ensures that no leaf reached from each node can have a cost less than the value of the bound function at that node.

As soon as one leaf has been found which has a cost equal to or less than the bounds at all the nodes elsewhere in the tree from which no further branches have been explored then that leaf corresponds to a vector with an optimum cost. A branching strategy is a set of rules for selecting which node to explore next by forming its branches. The most commonly used rules are:

Explore the node with the least bound

    (1) of all those so far unexplored,

    (2) of all those at the lowest level (i.e. furthest from the root),

    (3) of all those at the levels nearest the leaves which would spring from them.

In some problems (e.g. those on permutations) strategies 2 and 3 are the same.

Other more complicated strategies might be adopted, perhaps taking into account any difference in quality of the bounds at the different levels, or the amount of work involved in calculating those bounds. Different strategies can result in quite different amounts of work and so too can different rules for breaking any ties in the values of the bounds; unfortunately the effects can vary from problem to problem and also from one set of data to another in the same problem, and intuition can be an unreliable guide in picking which rules lead to least work. General advice must therefore be treated with caution but it is obvious that the number of nodes in the complete tree for any problem is greater even than the number of vectors (i. e. leaves in the tree) in a complete enumeration. It does not require much imagination to envisage a bad example in which strategy 1 failed to eliminate many branches of the tree; not merely would the work be heavy, but so too would the storage requirements to keep all the details of all the nodes from which branches would have to be examined. By contrast, strategies 2 and 3 tend to explore one limb of the tree at a time and to obtain quite quickly a complete vector with a best cost so far; these strategies enable some, and often many, nodes to be abandoned both during the progress to the leaf and during the backtracking.

In certain problems, for example those involving permutations, strategies 2 and 3 can be guaranteed not to have storage requirements which grow beyond a convenient size. In a tree for the permutations of n elements, there are n nodes at the first level, each of these nodes has $(n-1)$ branches, thus giving $n(n-1)$ nodes at the second level, and so on. The n bounds at the first level are computed and arranged in decreasing order; the smallest determines the branches to explore while the other $(n-1)$ are stored for later consideration. Next the $(n-1)$ bounds for the nodes explored at the second level are computed, ordered and $(n-2)$ of them retained; and so on. As the nth level is reached and the cost, C of the single complete permutation is calculated there are

$$(n-1) + (n-2) + \ldots + 3+2+1 = n(n-1)/2$$

other bounds are stored.

The one bound at the $(n-1)$th level is compared with the cost C ; if it is less, the complete permutation springing from it is formed, costed and compared with C , retaining the lesser, while if the bound is not less than C the backtrack goes to the smaller of the bounds at the $(n-2)$th level, eliminating the information on the lower level nodes. This elimination provides at least enough space for the bounds at the nodes to be explored before the next backtracking occurs. A diagram of the storage layout required is shown in figure 6. 4 and the corresponding portion of the tree for $n = 4$ is given in figure 6. 5, in both of which it is supposed that the leftmost nodes all happen to be the smallest at their levels.

B_2
B_3
.
.
.
B_n

Bounds at the (n-1) nodes
at level 1
$B_2 \le B_3 \le \ldots \le B_n$

B_{13}
B_{14}
.
.
.
B_{1n}
.
.
.
$B_{12..n-2,n}$

Bounds at the (n-2) nodes
at level 2 branching from
the node (1) with best bound
at level 1

Bound at the node at level (n-1)

Fig. 6.4. Storage for a permutation tree

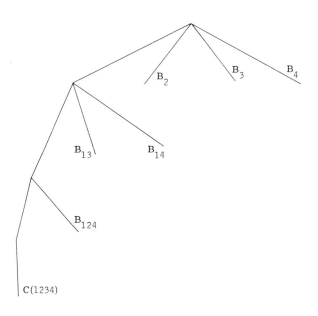

Fig. 6.5. A portion of the permutation tree

A convenient form of storage organisation is a stack at which the entries and exits are made at the end corresponding to the highest level node (i.e. at the foot of figure 6.4). When the stack becomes empty an optimum has been found and proved to be so.

6.2.3 A Scheduling Example

The descriptions of the last few paragraphs are best illustrated by an example small enough to be worked completely with pencil and paper. Suppose that we have four jobs to do, that each job requires a machine A for a known length of time, then a machine B for another time and finally machine C for another period. The machines can only work on one job at a time and the jobs must go through the machines in the sequence A, B, C - no jumping to another machine, or interrupting one job to process another is allowed. The task is to find an order of doing the jobs that gets them all completed as quickly as possible. The name 'job shop scheduling' is usually given to this problem in the computing and operations research literature.

For the particular example to be worked the times are shown in table 6.1. We must, of course, be able to work out the time of completion

Table 6.1   Machine times required

Job	Machine		
	A	B	C
1	3	4	5
2	7	2	1
3	6	5	3
4	2	5	7

for the jobs for any given order of doing them, and we shall need bounds for the completion time given the order for the first one, two or three jobs. For pencil working, the time of completion is conveniently computed by a table showing the times at which the several machines are free to take another job.

For example, consider the jobs processed in the order 1, 2, 3, 4.

Table 6.2   Times machines become free

Job	Machine		
	A	B	C
1	3	7	12
2	10	12	13
3	16	21	24
4	18	26	33

Table 6.2 is completed row by row. Job 1 leaves machine A at time 3,
leaves B at 7 and C at 12. Job 2 goes on to A at 3 and leaves it
at 10, and as B is free (since 10 > 7) then it goes straight on to it,
leaving at 12, and so to C (which has just become free) which it leaves
at 13. Similarly for jobs 3 and 4; it happens that job 4 leaves A at 18
but then has to wait until 21 before it can start its work on B. The
entry in column B is formed from the sum of the time required on B
and the greater of the current entry in A and the previous entry in B;
and column C is constructed similarly.

A simple bound at the first level follows from the observation that
machine B starts working when the first job reaches it; it can do no
better than work continuously for the sum of the times on B and leave C
to complete the job (other than the first one) which needs the shortest
time on it. Thus a bound is:

$$a_i + \Sigma b + \underset{j \neq i}{\text{Min}}\ c_j \tag{6.1}$$

where job i is processed first. Similarly a bound at the second level is:

$$a_i + a_j + (\Sigma b - b_i) + \underset{k \neq i, j}{\text{Min}}\ c_k \tag{6.2}$$

where jobs i, j are processed first and second respectively; we shall see
that this bound is not as tight as (6.1) always because machine B will
not necessarily be free to start job j immediately it leaves A. (Some
other bounds are readily available in this problem which are mentioned
in section 6.2.4.) In such a case the bound at the node above is preferred.
These cases are shown on the tree in figure 6.6 with this bound in square
brackets [ ]. A similar bound at the next level is

$$a_i + a_j + a_k + (\Sigma b - b_i - b_j) + c_l \tag{6.3}$$

where i, j, k are the first three jobs in that order, and l is the last job.

It would be possible to find better bounds than these but as they
would take a little more computing we shall use (6.1), (6.2) and (6.3) to
illustrate the processes. The numbers at the nodes in figure 6.6 are
these bounds while the orders of the jobs represented by the nodes are

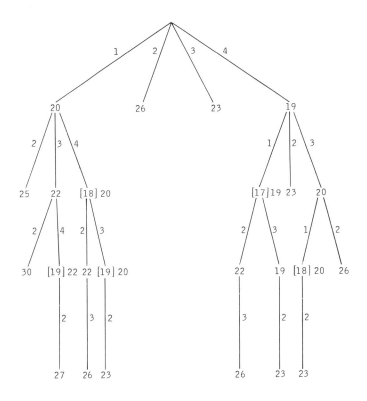

Fig. 6. 6. Branch and bound for a scheduling problem

obtained by the sequence of numbers on the branches leading to the
nodes; thus the rightmost leaf of the tree corresponds to the order
(4, 3, 1, 2) of the jobs and needs a time of 23 units.

Strategies 2 and 3 are equivalent for this problem as the paths to
all leaves of the tree are of the same length. First the four bounds 20,
26, 23, 19 at the first level are computed; the least bound is 19 at
node 4 . Then the three bounds [17], 23, 20 are computed at the nodes
below node 4 at the second level, i. e. (4, 1), (4, 2), and (4, 3) . The
bound calculated from (6. 2) at node (4, 1) is less than the bound 19 at
the node above it and cannot therefore be attained; accordingly, we take
the bound at node (4, 1) to be 19 again and as it is the least of the three
we proceed to calculate the bounds from (6. 3) at the two nodes below
(22, 19) . The completion time for the job order (4, 1, 3, 2) is then
found to be 23 .

At this stage we can see from the tree that no order starting with job 2 can complete as quickly, and that no order starting with job 3 can do better (although it is possible that one could do as well). This information would not be immediately available if the stack storage organisation described in section 6.2.2 had been adopted; in this case, as well as in the pencil solution we notice that there is a bound of 22 at node (4, 1, 2) which might improve the time so we backtrack there and branch to the leaf (4, 1, 2, 3) but find that its time for completion is 26 .

Backtracking now takes us to the second level where node (4, 3) has a bound of 20 ; its branches lead to a bound [18] which is unattainable and so is left at 20 , and to one of 26 which is of no further interest. The complete job order (4, 3, 1, 2) is formed and found to complete in the same time as the previous best, 23 .

Backtracking again, the only unexplored node is (4, 2) at the second level but as its bound is 23 , no leaves below it can improve on the best job orders already found. Further backtracking leads to node 1 where the bound is 20 ; branching from it starts exploration from node (1, 4) ; and thence to node (1, 4, 3) and the order (1, 4, 3, 2) which has also a completion time of 23 units.

Node (1, 4, 2) must be explored to see if its bound, 22 , can be attained but it is discovered that the order (1, 4, 2, 3) has a time 26 . Backtracking to node (1, 3) still shows a possibility of improving on a time of 23 so its branches are explored, giving bounds [19] (unattainable) and 30 . The order (1, 3, 4, 2) is found to need 27 units to complete.

The remainder of the backtracking is necessary to demonstrate that the time 23 is indeed the best possible. The bound 25 at node (1, 2) is now at the top of the stack and is discarded. Next the bound 23 at node 3 need not be explored unless all the orders which have the best time have to be found. Finally the bound 26 at node 2 is discarded, and the search is complete.

The stack initially fills as shown in table 6.3 in the process of obtaining the first complete job order. Each node with a bound less than the best completion time so far found has to be explored as the stack releases its data from its lower end. When it starts to fill again after backtracking it is unnecessary to insert into the stack any nodes with bounds which offer no prospect of improvement. For example, when the backtrack reaches node 1 , the branch to node 1, 2 can be discarded as its bound is not less than 23 .

Table 6. 3   The stack of bounds and nodes

Node	Bound
2	26
3	23
1	20
(4, 2)	23
(4, 3)	20
(4, 1, 2)	22

The reader should check that in this example strategy 1 would explore the nodes in a rather different sequence and one needs to adopt some rule for breaking ties between bounds - for example 'If the bounds are the same, explore the node nearest a leaf, or the leftmost node if the levels are the same. ' The same route is followed to the first complete order $(4, 1, 3, 2)$ but when this is reached there are two nodes with bounds of 20 $((4, 3)$ and 1) ; the former will be explored next but both they and nodes $(4, 3, 1)$, $(4, 3, 1, 2)$, $(1, 4)$ and $(1, 4, 3)$ will all be explored before node $(4, 1, 2)$ which came early in the backtracking in strategy 2.

6. 2. 4 Discussion

In the example of section 6. 2. 3, each of the strategies suggested finds an optimum at the first leaf reached; this is rather lucky but it is not uncommon to find quite early in the search what later turns out to be an optimum solution when the bounding functions are reasonably good. Unfortunately a great deal of the work in branch and bound searches is often solely concerned with verifying that a solution obtained early really is optimum. For example, in one job scheduling problem with 8 jobs, an optimum was found after considering 58 nodes but it was not proved until 8097 nodes had been examined, while in a 9 job problem the first trip down the tree found an order that had not been improved upon 15 000 nodes later, when the search was abandoned.

This work (or some of it) can be avoided if one is prepared to settle for a guarantee that a solution that has been obtained is certainly 'nearly' optimum even if it turns out not to be optimum - an assurance that no one else will find a solution that is much better than the one we have. In our example, when the job order $(4, 1, 3, 2)$ was reached it was known from the bounds already calculated that no other order could have a completion time below 20 units; the improvement possible was therefore at most 3 units; if such a gain would not be significant in the practical application the search could stop. A search for an improvement of more than one unit would just avoid exploring nodes $(4, 1, 2)$ and $(1, 3)$. In general a relaxation of the target of the search cannot increase the work required and will usually decrease it, perhaps substantially, but forecasts of how much are rarely available.

The quality of the bounds can also affect the amount of work in a calculation. Sometimes it is difficult enough to discover just one bound that can be used but in other problems there will be a choice between a simple bound and one which is better but rather more complicated; it will be a matter of intuition or even guesswork which choice will give the optimum most economically. For example, in the 'job shop scheduling' problem there are several quite simple bounds to choose from. If a node represents the partial schedule of jobs (i. e. partial permutation) Π_r then a lower bound for the completion time is

$$\text{TIME A } (\Pi_r) + (\Sigma a_i - \underset{\overline{\Pi}_r}{\Sigma} a_i) + \underset{\overline{\Pi}_r}{\text{Min}} (b_i + c_i) \qquad (6.4)$$

which is the time to complete the partial schedule on machine A , followed by the rest of the processing on that machine together with the time of the job not in the partial schedule (denoted by $\overline{\Pi}_r$ in equation (6.4)) that can complete most quickly on the two other machines. There is a similar bound based on the completion of work on the second machine. Bound (6.4) gives for the first level values 21, 26, 21, 21 which are better twice, and worse once, than (6.1), but which lead to more work. Yet other simple bounds exist:

$$\Sigma b + \text{Min}(a_i + c_j) \qquad (6.5)$$

and

$$\text{Min}(a_i + b_i) + \Sigma c \qquad (6.6)$$

148

Of these (6.6) is, in our example, one which eliminates half the tree at the first level once a leaf with time 23 is obtained. Unfortunately there are examples where better bounds lead to more work in the exploration of the tree so that while good bounds usually help considerably this cannot be guaranteed.

6.3 DYNAMIC PROGRAMMING

6.3.1 Concepts

Many problems of searching for an optimum can be considered in several stages, with a choice to be made at each stage; the choices made will determine whether an optimum is reached. In general the number of stages could be infinite - a choice whether to fire the engines of a rocket might be theoretically possible at any point in its motion - but in the context of this book we concentrate upon finite discrete situations. One approach which could be tried to solve such problems is to consider all possible combinations of the choices at every stage but it is obvious that the number of combinations can soon make the computations infeasible as the size of the problem - the number of choices and number of stages - grows. The method which has been given the name 'dynamic programming' looks at the problem differently. It recognises that once choices have been made for any of the stages there is only a possibility of reaching an optimum if the choices in the remaining stages are made in the best possible manner - if the choices were not so made, clearly the best overall outcome may not be obtained. This 'principle of optimality' - that however many stages are left they must be completed in an optimum way - enables an optimisation problem with n stages (and so n-dimensional in a certain sense) to be tackled as a number of one-dimensional optimisation problems. This simplification can be of great help in some problems but in others the number of one-dimensional problems, even when they are of a similar type, which have to be solved and stored in the course of the computation grows so rapidly and imposes severe limitations on the size of problem that can be tackled.

It will be evident that the word 'programming' of dynamic programming has nothing at all to do with writing instructions in a computer language; its derivation springs from the idea of forming a programme (the spelling '-me' in Queen's English emphasises the difference) of action when the choices at each stage present themselves, and it is thought 'dynamic' because of the progression through the stages and the dependance of the current choice on the preceding ones.

As it has been so far presented, dynamic programming would seem to be applicable only to those problems which naturally embody choices at a number of stages; in practice, however, problems which can be indexed by some integer variable, n say, can often be tackled by the method of thinking of the given problem for the particular value of n as being embedded in a sequence of problems for varying values of n, each member of the sequence being one of the stages. For example, the 'travelling salesman' problem of finding the shortest tour which visits each of 10 cities once and once only is treated as the special case $n = 10$ of the general n-city problem and the dynamic programming approach considers the problems for the sequence of values $n = 1, 2, \ldots$

The dynamic programming formulation of an n-stage optimisation focuses attention on the value of the optimum and not on the way in which the optimum is obtained (which is derived incidentally during the computations). The key step is to notice that the value of the optimum depends on:

(1) the conditions and constraints of the problem,

(2) the number of stages, N,

(3) the initial values of the variables in the problem.

The value of the optimum can thus be written $F_N(\mathbf{X}_0)$ where $\mathbf{X}_0$ is a vector giving the initial values of the variables and the form of the function F is almost always unknown and often complicated. In the first stage of the problem a choice of the available variables, $\mathbf{X}_1$, will be made, and as a result of that choice the initial conditions for the next stage will have been changed from $\mathbf{X}_0$ to $\mathbf{X}_1$, say. In the examples that are presented here there will be a simple relationship - for example, an additive or multiplicative one - between the optimum values in the N-stage and (N-1)-stage problems. If the one-stage problems can be solved, their results can be used to solve the two-stage ones, and so to progress to three and four ... and higher stages.

6.3.2 A Simple Example

Consider the problem of partitioning a positive quantity x, into positive quantities $x_1, x_2, \ldots, x_n$ so that the product $x_1 x_2 \ldots x_n$ is as large as possible. Of course, if the x is a continuous variable an elementary calculus solution is available but a dynamic programming solution could be obtained as follows.

The maximum value of the product depends only on the quantity to be divided, x, and the number of pieces, n; let it be $f_n(x)$. Then

$$f_n(x) = \max_{0 \le x_n \le x} \{x_n f_{n-1}(x-x_n)\}$$

since as soon as one piece, x_n, has been chosen the remainder, $x-x_n$, must be divided into $(n-1)$ pieces to give the greatest product (by the Principle of Optimality) and the best of these x_n must be selected.

It is obvious that $f_1(x) = x$ for all x and it is a problem in one variable to establish that $f_2(x) = x^2/4$, which is obtained when $x_1 = x_2 = x/2$. There is a succession of one variable problems to find that $f_3(x) = (x/3)^3$, $f_4(x) = (x/4)^4$, and so on.

In this solution it should be noticed that after $f_1(x)$ has been found and used to obtain $f_2(x)$ it is not required again although the division which gives the optimum is needed to exhibit the method of obtaining it. Thus when a discrete computational solution is considered the storage required for function values at a set of points can be limited to that necessary for two stages at a time, the earlier ones being discarded (even though they may have to be regenerated later to find the route by which the optimum may be obtained).

If the variables of the above problems take only integer values the computational approach is the same as that adopted for the discrete approximation to the continuous problem. Consider the maximisation of the product $x_1 x_2 x_3$ where x_1, x_2, x_3 are positive integers whose total is 14. Let $f_r(N)$ be the maximum value of the product of r positive integers whose total is N. Clearly $f_1(N) = N$, so that we have the first row of table 6.4.

Table 6.4   Values of $f_r(N)$

										N					
r	0	1	2	3	4	5	6	7	8	9	10	11	12	13	14
1	0	1	2	3	4	5	6	7	8	9	10	11	12	13	14
2	0	0	1	2	4	6	9	12	16	20	25	30	36	42	49
3	0	0	0	1	2	4	8	12	18	27	36	48	64	80	100

The second row is completed using the relation

$$f_2(N) = \underset{0 < x < N}{\text{Max}} \{x f_1(N-x)\}$$

Thus

$$f_2(5) = \text{Max}\{f_1(4), 2f_1(3), 3f_1(2), 4f_1(1)\} = 6$$

If we are only interested in $f_3(14)$ we can proceed immediately to compute it, noting that its value, 100, is obtained by using either of the underlined values of $f_2(N)$, i.e. $f_2(9)$ or $f_2(10)$, showing that one of the x's for the optimum division is either 5 or 4. A similar expression of the $f_1(x)$ value which enters into the calculation of $f_1(9)$ or $f_2(10)$ gives the complete partition (4 5 5) of 14 to obtain the maximum.

Table 6.4 shows the complete row for $f_3(N)$ which would be required to compute $f_4(14) = 144$.

6.3.3 Paths through a Network

In the following artificial little problem the task is to find a path in figure 6.7 starting at any of the points on line A and finishing on line B which has least sum of weights attached to the arcs of the path.

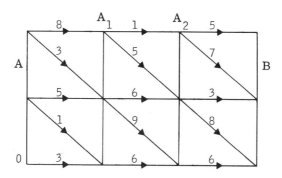

Fig. 6.7. A network example

There is a steady progression from left to right so that we can regard the intermediate vertical lines, marked A_1, A_2 as stages in the problem. In this case there are always three stages but more complicated networks could have larger and variable numbers of stages. If there are coordinate axes with origin 0 we can define a function $f(i, j)$ as the least total weight of a path from the point (i, j) to the line B. We need to find $\underset{j=0,1,2}{\text{Min}} \{f(0, j)\}$, and the corresponding path.

Let $n(i, j; k, l)$ be the weight of the arc (if any) from (i, j) to (k, l).

Then

$$f(i, j) = \text{Min} \{n(i, j; i+1, j)+f(i+1, j), \ n(i, j; i+1, j-1)+f(i+1, j-1)\}$$

Thus we can build up table 6.5 below, which also shows the route by which the various minima are reached (H = horizontally, D = diagonally).

Table 6.5　Best paths

y	A_2:2	A_1:1		A:0	
2	5 H	Min(1+5, 5+3) = 6	H	Min(8+6, 3+9) = 12	D
1	3 H	Min(6+3, 9+6) = 9	H	Min(5+9, 1+12) = 13	D
0	6 H	6+6 = 12	H	3+12 = 15	H

Thus 12 is the weight of the best path and the route D, H, H; i.e. $(0, 2), (1, 1), (2, 1), (3, 1)$.

6.3.4 Permutation Problems

Optimisation in permutation problems can usually be tackled by dynamic programming but, in common with other methods, the amount of computation often grows rapidly with the size of the problem. The approach can be illustrated in the 'travelling salesman' problem; suppose we wish to find the shortest route from a city 0, through each of cities $1, 2, \ldots, n$ once and once only, returning to city 0. Let $f(i:j_1, j_2, \ldots, j_r)$ be the shortest route from city i, through the cities $j_1, \ldots, j_r$ to city 0. The shortest route in the above version of the 'travelling salesman' problem is $f(0:1, 2, \ldots, n)$. The principle of optimality gives a relation between functions $f(.)$ with different numbers of arguments and the distances $d(i, j)$ between cities i and j; if one moves first from city i to city j_k the shortest route starting from i in that way, which passes through $j_1, \ldots, j_r$ and ends at 0 has length

$$d(i, j_k) + f(j_k:j_1, j_2, \ldots, j_{k-1}, j_{k+1}, \ldots, j_r)$$

Accordingly

$$f(i{:}j_1, \ldots, j_r) = \min_{1 \le k \le r} \{ d(i, j_k) + f(j_k{:}j_1, \ldots, j_{k-1}, j_{k+1}, \ldots, j_r) \}$$

The computation proceeds by obtaining

$$f(i{:}j) = d(i, j) + d(j, 0)$$

for all pairs of cities (i, j) where $i, j \in \{1, \ldots, n\}$; next all the $f(i{:}j, k)$ are found and so on. Clearly the storage needed grows quickly (perhaps of less practical importance now that the cost of computer memory has fallen) as well as the computing time (cf. exercise [6.18]).

6.4 COMPLEXITY

Already in this book and in others the student of computing science will have encountered different algorithms for the same problem and will have compared the computing resources which they require: for example, the numbers of additions, multiplications, comparisons, storage references and storage locations. The impression will have been gained that one algorithm is 'better' than another for a given problem in certain circumstances, perhaps in all circumstances. It is natural to hope to find a 'best' algorithm for a problem for a useful range of circumstances but even if such a quest proves fruitless an idea may have been gained of how good a 'best' algorithm for the problem would be if only one could be found. Thus an idea is formed of the amount of computing that a problem can need and, as a next stage, the relative demands of general cases of two or more problems. It is this concept that the study of computational complexity formalises; it focusses attention upon the problems themselves - for example, sorting, the 'travelling salesman' problem or the inversion of a matrix - rather than on specific algorithms for solving them, and attempts to assess the number of steps that would be required by an idealised computing machine to solve any instance of a given problem as a function of its size - the number of items to be sorted, the number n of cities to be visited or the number n of rows in the matrix to be inverted. The size of some problems may depend on several parameters; in job shop scheduling, the number n of jobs, the number m of machines together with any constraints on the precedence of the jobs all influence the size of the problem. For some purposes in complexity theory it is convenient to reduce them to one parameter (for example, the number of bits necessary to express these characteristics) so that one can study how the number of steps required varies as a single size parameter increases.

The problems which have so far attracted most study by theorists working in complexity fall into two classes. The first is composed of those problems for which the number of steps, i. e. the complexity, is of the order of a polynomial in the size. Examples are some operations on matrices of order n and the evaluation of polynomials of degree n where the complexity is $O(n^k)$ for some constant k. Such problems are generally tractable on modern computers for quite large values of n, efficient algorithms for producing an exact solution may exist for them, some having emerged from the theoretical studies of complexity.

The other class contains some of the problems encountered in this book, like some 'travelling salesman' and 'job shop scheduling' problems. The situation with these problems is not yet completely clear; each problem can be transformed into one of the others in the class in a number of steps which is polynomial in the size, so that if one problem has polynomial complexity, they all have; while if one can be shown to be of higher than polynomial complexity, then all have been shown to be non-polynomial.

What has been proved is that problems in this class, called NP, can be solved in 'polynomial time' (i. e. a number of steps which is a polynomial in the size of the problem) on a 'non-deterministic' computer. Such an abstract machine has no counterpart in reality as its special feature is an operation which makes another copy of the computer which also proceeds through the computation; thus after r occurrences of this special operation there would be 2^r copies of the computer in existence. While this theoretical result suggests that problems in the class NP are likely to require exponential time on a real computer no one has yet managed to prove (or disprove) it. A problem is said to be NP-complete if every problem in the class NP can be reduced to it in polynomial time.

As the search for a polynomial time algorithm for these problems has been long and vigorous by many skilled people, and has brought no better result than an exponential time one, the practical conclusion to draw is that NP-complete problems can not be solved exactly on present computers unless the problems are small in size. Bigger problems can only be tackled approximately and some effort should be invested in finding good heuristic approaches.

6.5   BIBLIOGRAPHY

Different methods of exhaustive enumeration closely related to the backtrack procedure described here have been applied many times and reported in the literature. Early general expositions of backtrack were given by Walker [1], Lehmer [2] and Golomb and Baumert [3]. Walker was the first author to use the apt name 'backtrack'; Lehmer's brief account of the technique is very readable whilst the very good general formulation of Golomb and Baumert contains applications to the Queens' problem, comma-free codes, systems of distinct representatives and sum-free sets. More recently Wells [4], who prefers the term 'tree programming', has examined backtrack in some depth.

The reader has been warned that some enumerations can be long because they are large, but backtrack programs can be lengthy even if the set of accepted vectors is not large; Knuth [5] has described a method which helps to predict the work involved.

Branch and bound methods have been surveyed by Lawler and Wood [6] and an introduction to the 'job shop scheduling' problem is given by Sisson [7]. Bellman has had a great deal to do with the development and popular exposition of dynamic programming; he has written many books and articles on the subject (see for example [8]). The relationship between branch-and-bound and dynamic programming is demonstrated by means of a sample problem in Kohler and Steighitz [9]. Finally Aho, Hopcroft and Ullman [10] contains a good discussion of complexity and NP-complete problems as well as an application of dynamic programming to the multiplication of matrices.

[1]   R. J. Walker: 'An Enumerative Technique for a Class of Combinatorial Problems. ' pp91-4 in 'Procs. Symposia in Appl.

Math. Vol 10. Combinatorial Analysis, ' edited by R. Bellman and M. Hall Jr. American Math. Soc. Providence R. I. , 1960.

[2]   D. H. Lehmer: 'The Machine Tools of Combinatorics. ' Chapter 1, pp5-31 in 'Applied Combinatorial Mathematics' edited by Beckenbach, John Wiley, 1964.

[3]   S. W. Golomb and L. D. Baumert: 'Backtrack Programming. ' J. Ass. Comp. Mach. , Vol 12 (1965), pp516-24.

[4]   M. B. Wells: 'Elements of Combinatorial Computing. ' Chapter 4, pp93-126. Pergamon Press, 1971.

[5]   D. E. Knuth: 'Estimating the Efficiency of Backtrack Programs. ' Mathematics of Computation, Vol 29 (1975), pp121-36.

[6]   E. L. Lawler and D. E. Wood: 'Branch-and-bound Methods: A Survey. ' Operations Research, Vol 14 (1966), pp699-719.

[7]   R. L. Sisson: 'Sequencing in Job Shops - A Review. ' Operations Research, Vol 7 (1959), pp10-29.

[8]   R. E. Bellman: 'Dynamic Programming. ' Princeton University Press, 1959.

[9]   W. H. Kohler and K. Steiglitz: 'Enumerative and Iterative Computational Approaches. ' Chapter 6, pp229-87 in 'Computer and Job-shop Scheduling Theory' edited by E. G. Coffman Jr. John Wiley, 1976.

[10] A. V. Aho, J. E. Hopcroft and J. D. Ullman: 'The Design and Analysis of Computer Algorithms. ' Addison-Wesley, 1974.

6. 6   EXERCISES

[6. 1] Generate the compositions of the integer 5 by a backtrack process and draw the corresponding tree.

Similarly generate the partitions of 5 and draw the tree.

[6. 2] Generate by backtrack all the permutations of the first five integers which have no integer removed by more than one place from its position in the natural order (i. e. i may only occupy positions i-1, i, i+1).

[6. 3] Generate the first twenty permutations in lexicographical order of the first five integers which have no integer more than two units greater than both its neighbours (or than its single neighbour if at an end).   (cf. exercise [2. 15].)

[6.4] Describe the backtrack process for constructing all the vectors of a finite set satisfying given conditions.

Give an algorithm (specified by flow diagram or in an alternative precise manner) for constructing all valid Algol expressions using exactly k of the symbols 0, 1, 2, 3, 4, 5, 6, 7, 8, 9, +, -, ×, ↑. By considering the final one or two characters of such expressions (or otherwise) obtain a recurrence relation between the number of expressions of different lengths and hence find the number for length k.

(Newcastle 1969; cf. section 2.6)

[6.5] The following assignment problem can be solved by backtrack. The figures in the assignment matrix below represent the cost of man i doing job j. What assignment of men to jobs would give a minimum cost? What is the cost? Is there more than one minimum?

Man	Job			
	1	2	3	4
1	5	7	7	3
2	5	7	7	3
3	7	6	6	3
4	9	8	10	6

[6.6] The sequence $\{2, 3, 4, 2, 1, 3, 1, 4\}$ which contains the first 4 ($=n$) integers 2 ($=r$) times each, such that consecutive occurrences of each integer i are separated by i elements, is called an $L(2, 4)$ sequence.

(a) Outline an algorithm (by flow diagram or otherwise) for checking whether a given sequence is an $L(r, n)$ sequence.

(b) Suggest a method which might be used to derive all possible $L(r, n)$ sequences (or to check if none exists) and estimate for what values of r, n the computation would be feasible on a modern large-scale computing system.

(Newcastle 1972)

[6.7] It is known that every positive integer is the sum of eight or less numbers of the form $Q_n \equiv (n^3+5n)/6$. Describe the design of a program to display all the integers less than $100\,000$ which are not the sum of four or less of the Q_n. Comment on any storage problems which might be encountered and suggest methods of overcoming them.

How could the problem of displaying all the different partitions of integers into at most four parts of the form Q_n be programmed? (Detailed coding is not required; descriptions may be given by flow diagram, a language of Algol type, or otherwise as appropriate.)

(Newcastle 1970)

[6. 8] A combinatorial optimisation problem can be formally expressed as:

> Minimise $C(x)$ where x is a state vector of n dimensions where each element x_i can take integer values in the range $(0, X_i)$.

In these terms explain the meaning of:
- (a) lexicographic ordering,
- (b) degeneracy of optimal solution,
- (c) branch and bound,
- (d) backtrack,
- (e) heuristic approach.

Illustrate your answer by considering the problem of minimising the bandwidth of the 5×5 symmetric matrix by permuting rows and columns whose non-zero off-diagonal elements are in positions:

$$(1, 2), (1, 4), (1, 5), (2, 3), (2, 4), (3, 4), (4, 5)$$

(Newcastle 1972)

[6. 9] Describe the backtrack procedure for generating systematically the vectors of a finite set.

A pair of partitions with unequal parts of a positive integer, N, $(p_1 \ldots p_r)$ and $(q_1 \ldots q_s)$, is such that every integer i $(1 \le i < N)$ may be represented in one and only one way as a sum of either a subset of the (p_i) or a subset of the (q_i). For example, if $N = 9$ the partitions $(1\ 2\ 6)$, $(4\ 5)$ satisfy the conditions, but for $N = 12$, $(1\ 2\ 9)$, $(3\ 4\ 5)$ do not (since 3 and 9 can be represented twice and 6 not at all).

For given N , describe a backtrack procedure for finding all pairs of partitions which satisfy the above conditions, illustrating it by a suitable tree structure or otherwise.

Derive the partitions for $N = 17$.

(Newcastle 1970)

[6.10] In the application of the branch and bound strategy in which the bounds are kept and explored in order of size irrespective of the level to which they refer, find, for a problem dealing with permutations, the least number of nodes at which bounds have to be computed before the first leaf is reached, and the greatest number of nodes which might have to be retained during the proof that an optimum has been obtained.

[6.11] For the 'job shop scheduling' problem of section 6.2.3 perform the branch and bound search using strategy 1 and the bound (6.4).

[6.12] How many different subsets containing one or more different objects may be selected from n different objects?

Suggest a rule for ordering the subsets lexicographically and give the order for subsets of the objects $+, \times, -$.

A sequence $\{x_i, i = 1, 2, \dots, N\}$ of different positive integers x_i is given and it is desired to find the longest monotonic increasing subsequence $(\xi_1, \dots, \xi_r)$ that can be selected from the sequence while maintaining the original order of the integers in the sequence, i.e. we require $\xi_i < \xi_j$ for $i < j$ and if $\xi_i = x_m$, $\xi_j = x_n$, then $m < n$. One approach to this problem is to enumerate all such subsequences. Describe briefly (by flow diagram or otherwise) an algorithm for performing the enumeration. Suggest ways in which a more efficient solution of the problem might be found.

[6.13] A result similar to that quoted in exercise [6.7] holds for the tetrahedral numbers $T_n = (n^3 - n)/6$ and for $R_n = (n^3 - 7n)/6$. Write a program to display partitions of a given integer N into as few of the numbers T_n (or Q_n, R_n) as possible. (See H.E. Salzer and N. Levine, Math. Comp., Vol 22 (1968), pp. 191-2.)

[6.14] Write a program to accept an increasing sequence $\{i, j, k, \ldots\}$ and to output all subsets in increasing order of their totals.

[6.15] An 'up-down' permutation $a_1 \ldots a_n$ of $1, 2, \ldots, n$ is one in which $\mathrm{sgn}(a_i - a_{i+1}) = -\mathrm{sgn}(a_{i+1} - a_{i+2})$, where $\mathrm{sgn}(x)$ is -1, 0 or +1 according to whether x is negative, zero or positive. Devise a lexicographical ordering for these permutations and write a program to print them (16) for $n = 5$ and check that it would print 61 such permutations for $n = 6$. (See SIAM Review, Vol 10 (1968), p. 225.)

[6.16] A comma-free code is a collection of words, called the dictionary, from which sentences which do not contain any punctuation (i. e. no spaces, commas, etc.) may be formed without ambiguity. For example if the three-letter English words 'but' and 'one' are in the dictionary then 'ton' cannot be because of the ambiguity in 'butone'. Consider dictionaries consisting only of words with k letters - the letters can be chosen from an n-letter alphabet. In these cases the maximum possible number of words is n^k but most will not be comma-free. Write a backtrack program that will find the largest comma-free dictionary for a given k and n. Use it to show that, for $n = 2, k = 4$, the value is 3 ; what is the value for $n = 4, k = 4$? (See S. W. Golomb and L. D. Baumert [3].)

[6.17] Consider the problem of multiplying the n matrices

$$M = M_1 \times M_2 \times M_3 \times \ldots \times M_n$$

where matrix M_i has r_{i-1} rows and r_i columns. Using the normal method of matrix multiplication, so that multiplying a $p \times q$ matrix by a $q \times r$ matrix requires pqr operations, in what order should we multiply the matrices together so that the minimum number of operations is performed? For example take

$$M = \underset{(2\times5)}{M_1} \times \underset{(5\times10)}{M_2} \times \underset{(10\times1)}{M_3} \times \underset{(1\times20)}{M_4}$$

then if we multiply in the order $M_1 \times (M_2 \times (M_3 \times M_4))$ the number of operations is 1400 whilst in the order $(M_1 \times (M_2 \times M_3)) \times M_4$ it is only 100.

Trying all the orderings is impractical for large n so it is suggested that a dynamic programming approach to the problem should be used. If

$$m_{ij} = M_i \times M_{i+1} \times \ldots \times M_{j-1} \times M_j \quad (1 \le i \le j \le n)$$

then this approach calculates the m_{ij} in order of increasing difference between i and j. We start by calculating m_{ii} for all i, then $m_{i,i+1}$ for all i, then $m_{i,i+2}$ for all i, etc. At each stage when we calculate m_{ij}, the intermediate terms m_{ik} and $m_{k+1,j}$ are already available. Show how your algorithm will operate with the following data

$$n = 5, \ r_0 = 5, \ r_1 = 12, \ r_2 = 10, \ r_3 = 15, \ r_4 = 4, \ r_5 = 6$$

and give the resulting order of multiplication of these five matrices.

[6.18] A salesman is required to visit once and once only each of n different cities (numbered $1, 2, \ldots, n$) starting from a base city (number 0) and returning to this base city. Show how the method of dynamic programming may be applied to finding the path which minimises the total distance travelled. The function $f(i; j_1, j_2, \ldots, j_k)$, defined as the minimum of the lengths of paths from city i to the base city 0 which pass once and once only through each of the cities $j_1, j_2, \ldots, j_k$, may be used in the method if desired.

Estimate the storage requirements of this method for the problem with n cities to be visited from the base city and state what size of problem could be conveniently solved on a computer with 4000 words of internal memory. (Newcastle 1963)
(This question betrays the size of machine Newcastle workers had available in 1963 - something only a little larger than a pocket programmable calculator available in 1978 for less than £100! But

note that a machine with a thousand times more memory, 4 million words, can only contain the working for a problem about twice as large by this method.)

7 · Theorems and Algorithms for Selection

7.1 INTRODUCTION

There are many problems where we have to select objects given various restrictions on the selection; some arose in chapter 4 which examined permutations, combinations and distributions. We shall look at further examples of such problems in this chapter. A typical problem is to find an assignment of a number of applicants to a (possibly different) number of jobs which is in some sense optimal and which observes various conditions such as the inability of some applicants to do some of the jobs. Within this rather loose framework there are many types of assignment problems. Alternative names for such problems are matchings (used mainly in graph theory) or marriages. Assignment problems can usually be formulated in several different ways using set theory, graphs, trees or matrices. In this chapter the set description is preferred. We will discuss first a theorem in set theory on distinct representatives which leads to Konig's theorem on matrices of zeros and ones and to the classical assignment problem.

In the classical assignment problem there is a cost matrix whose general element c_{ij} is the cost of the ith man doing the jth job. The solution to this assignment problem is obtained when each man is assigned a different job such that the sum of the costs of the assignment is a minimum; there is a corresponding problem seeking a maximum of the total profits. An alternative assignment problem involves 'choice' matrices where the men rank the jobs in the order they would prefer to do them. This introduces new ideas of optimality and we examine one such criterion known as 'stable marriage'.

7.2   SYSTEMS OF DISTINCT REPRESENTATIVES

Before we prove the basic theorem on distinct representatives (due to Philip Hall) we will illustrate the ideas by an example.   Five committees contain various MP's as their members; it is decided to form a delegation of 'distinct representatives' with a nominated representative from each committee so that no MP represents more than one committee.   Let the MP's be labelled A, B, C, D, ...   and consider the following committees:

$$C_1 = \{A, B, C\} \qquad\qquad C_2 = \{A, B\}$$
$$C_3 = \{A, C, D, E, F\} \qquad C_4 = \{C, D, E\}$$
$$C_5 = \{C, D, E, F\}$$

Then we can choose as the distinct representatives A, B, C, D and E respectively from the five committees.   However if the committees are formed as follows:

$$C_1 = \{A, B\} \qquad\qquad C_2 = \{A, B, C, D, E\}$$
$$C_3 = \{B, C\} \qquad\qquad C_4 = \{A, B, C\}$$
$$C_5 = \{A, C\}$$

there is no system of distinct representatives.   This result can easily be seen by noticing that the four sets C_1, C_3, C_4 and C_5 require four distinct representatives and only contain the three MP's A, B and C between them.

In the generalisation of the problem to n sets $S_1, S_2, \ldots, S_n$ we wish to find a representative $x_i \in S_i$ for each set $(i = 1, 2, \ldots, n)$ such that all the x_i are different.   The problem is in essence in two parts: first, does a system of distinct representatives exist, and second, if it does, find such a system.   The basic theorem proved below answers both parts of the problem because the constructive proof gives as its result a system of distinct representatives if one exists.

Theorem (Philip Hall)

Given n finite sets $S_1, S_2, \ldots, S_n$ a necessary and sufficient condition for selecting a set of distinct representatives is that condition C never occurs.

Condition C. No k ($\leq n$) sets exist which between them contain less than k distinct members.

Proof. (Necessity) If there exists k sets $S_{i_1}, \ldots, S_{i_k}$ which between them contain only $k-1$ distinct elements then no set of distinct representatives can exist because at least one set must be unrepresented.

(Sufficiency) Suppose that condition C never occurs. Choose an arbitrary element x_1 to represent S_1 and continue to choose representatives for sets $S_2, S_3, \ldots$ as long as possible such that the x_i chosen for S_i is different from all the previous $i-1$ representatives. If we can continue this process up to and including the final set S_n then we have found a set of distinct representatives and proved the theorem.

Alternatively we reach some set S_j (say) all of whose elements $b_1, b_2, \ldots, b_t$ have already been used as representatives of the previous $j-1$ sets. Construct a list, called the B-list, with elements $b_1, b_2, b_3, \ldots,$ the first t elements of which are the elements $b_1, b_2, \ldots, b_t$ of set S_j. These elements are taken as the roots of the trees $T_1, T_2, \ldots, T_t$ which are constructed as follows:

Starting at $i = 1$, for each tree T_i with root b_i, we examine the set for which b_i is the representative; if any elements of this set are not in the B-list then add them to it and put them on the tree T_i with b_i as their father. Now the nodes at level 2, level 3, etc. in tree T_i are examined. Examining a node x consists of two parts. (i) Test to see if it is being used as a representative; if it is, examine the set it represents and add to the B-list and the current tree T_i any member of that set not already in the B-list. These members are put on the next level to x and linked to it. (ii) If x is not being used as a representative then it is used as a representative of the set at present represented by its father, the father is used to represent the set at present represented by his father and so on until we arrive at one of the roots $b_1, b_2, \ldots, b_t$ and the element released can be used to represent S_j.

There is one further case to consider: suppose that the trees $T_1, T_2, \ldots, T_t$ have been completely constructed and that all the elements of the B-list have been examined. In this case all these r elements in the B-list are representatives of sets and these sets only contain those r elements. In addition to this we have set S_j so we have $(r+1)$ sets with only r elements violating condition C of the theorem.

This proves the theorem since if condition C never arises we can continue finding distinct representatives until we complete all the n sets.

Examples

(i) Consider the sets

$$S_1 = \{1, 2, 3, 4\} \qquad S_5 = \{1, 8\}$$
$$S_2 = \{3, 5\} \qquad S_6 = \{6, 8\}$$
$$S_3 = \{3, 5, 7, 8\} \qquad S_7 = \{2, 3\}$$
$$S_4 = \{2, 5\} \qquad S_8 = \{2, 6\}$$

Take the first possible element to represent a set, then we have the representatives $1, 3, 5, 2, 8, 6$ for the first six sets. At this stage the elements of the set S_7 have both been used as representatives.

The trees are now constructed as follows:

$$\text{T}_1 \qquad\qquad \text{T}_2$$

$$2 \qquad\qquad 3 \qquad \text{B-list} = 2, 3, 5, 7, 8, \ldots$$

At this stage when we examine 7 we find it is not being used as a representative so it can be used to represent S_3 (releasing 5 which previously represented S_3); and 5 can now be used to represent S_4 (previously represented by 2) leaving 2 free to represent set S_7. Thus when we find an element (such as 7 above) which has not yet been used as a representative we can start a chain of moves going from the leaf to the root of the appropriate T-tree which allows us to release a representative for the blocked set.

We can now consider set S_8 but again both its elements are being used as representatives. We construct the following trees:

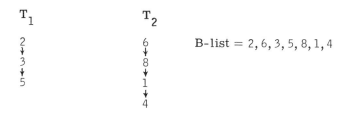

When we examine 4 we find it has not yet been used as a representative and therefore can be used to represent S_1 , releasing 1 to represent S_5 , 8 to represent S_6 and therefore 6 can be used to represent S_8 . So the first set of distinct representatives is $\{4, 3, 7, 5, 1, 8, 2, 6\}$.

(ii) Consider the sets

$R_1 = \{A, B, C, F\}$ $R_5 = \{C, D, E\}$
$R_2 = \{C, D\}$ $R_6 = \{A, F, G\}$
$R_3 = \{C, D, E\}$ $R_7 = \{A, D, E\}$
$R_4 = \{A, C\}$

Select representatives A, C, D starting at set R_1 for the first three sets. Both R_4's elements are used as representatives so set up the trees

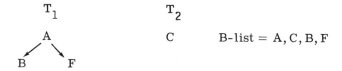

Use B to represent R_1 and A to represent R_4 . Continuing the process E can be used to represent R_5 and F to represent R_6 . However all R_7's elements are used as representatives which are $\{B, C, D, A, E, F\}$. Set up the trees

T_1 T_2 T_3

A   D   E   B-list $= A, D, E, C$
$\downarrow$
C

All the elements in the B-list have been examined; they represent sets R_2, R_3, R_4, R_5 . These sets together with R_7 form a system of five sets with only four different elements which violates condition C , so that no set of distinct representatives exists.

7.3 ALGORITHMS FOR SYSTEMS OF DISTINCT REPRESENTATIVES

Two algorithms for finding systems of distinct representatives will be discussed in this section. The first, which we will call Hall's algorithm, is based on the theorem given in the previous section. The second algorithm uses basic backtrack principles and thus is based on the ideas introduced in section 6.1.

7.3.1 Hall's Algorithm

The detailed algorithm for Hall's method follows the constructive proof given in section 7.2. We try to select a representative for each set until we reach a set, S_i , in which all the elements have already been used as representatives of the previous sets. Then we set up the B-list and the forest of trees rooted on the elements of S_i . If the B-list is exhausted without finding an element to use as a distinct representative then we exit to the **procedure FAIL**. When a new representative is found we make the appropriate changes to the previous representatives and carry on with the set S_{i+1} . The variables and data structures used in the algorithm are:

a parameter n , the number of sets.

a two-dimensional integer array S such that $S(i, j)$ is the jth element in the ith set.

a one-dimensional integer array SDR which holds the current distinct representatives so that $SDR(i) = j$ denotes that the jth set is represented by i .

a second one-dimensional integer array members such that members (i) is the number of elements in the ith set.

a procedure FAIL - to which we exit if it has been shown that no system of distinct representatives exist.

The local variables and data structure used are:

X is the number of elements in the B-list and so is set initially to the number of elements in the set S_i and is incremented as the B-list gains members.

ptr is used to determine the element we are currently examining in the B-list.

Blist(1::n) is the list of elements we test for a new distinct representative.

mark(1::n) is a logical (boolean) array which determines whether or not an element is in the Blist so that mark(R) = **false** if the element R is in the B-list.

Father(1::n) holds the father of a node of the trees of the B-list. If Father(j) = 0 then Blist(j) is a root and thus an element in set i.

Link(1::n) is used when we are changing representatives. Thus Link(t) is used to say that the new representative is for the set in Link(t).

The detailed algorithm is given below.

```
procedure HALL (integer value n; integer array S(*, *);
                    integer array SDR, members (*); procedure FAIL);
begin integer X, ptr, k;
    logical array mark (1::n);
      integer array Blist, Link, Father (1::n);
    for j:=1 until n do SDR(j):=0;
    SDR(S(1, 1)):=1;
    for i:=2 until n do
    begin X:=members(i);
            for j:=1 until X do
            if SDR(S(i, j))=0 then
             begin SDR(S(i, j)); goto NEXTI
             end;
            comment we have now examined all the ith set and its
            elements are all being used as representatives.   So we set
            up the B-list and the trees;
            for j:=1 until n do mark(j):=true;
            for j:=1 until X do
            begin Blist(j):=S(i, j);
                mark(Blist(j)):=false;
                Father(j):=0;
            end;
```

170

```
    ptr:=1;
    while ptr≤X do
    begin if SDR(Blist(ptr))≠0 then
        begin k:=SDR(Blist(ptr));
            Link(ptr):=k;
            comment we now examine the kth set, which was represented
            by current element in the B-list, to see if there are any
            elements in it which are not in the B-list;
            for j:=1 until members(k) do
            if mark(S(k, j)) then
            begin comment add this element to the B-list;
                mark(S(k, j)):=false; X:=X+1;
                Blist(X):=S(k, j); Father(x):=ptr
            end;
            ptr:=ptr+1;
        end
    else begin comment a new representative has been found.  Change
                representatives so that set i can be represented by one
                of its elements;
                while Father(ptr)≠0 do
                begin SDR(Blist(ptr)):=Link(Father(ptr));
                    ptr:=Father(ptr);
                end;
                SDR(Blist(ptr)):=i; goto NEXTI;

        end
    end;
    FAIL; comment this is the case where no SDR can be found;
NEXTI:end iloop;
end HALL;
```

7.3.2 Backtrack Algorithm

This is a much more straightforward algorithm and it will provide
all the possible systems of distinct representatives. If we look at
example (i) in section 7.2, the method selects the following representatives

$$S_1\text{-}1, \; S_2\text{-}3, \; S_3\text{-}5, \; S_4\text{-}2, \; S_5\text{-}8, \; S_6\text{-}6$$

At this stage both the elements 2 and 3 of S_7 have been used as representatives so we backtrack to S_6 and try the next element in it, i. e. 8 . This element is also being used as a representative so we must backtrack again. We go back in fact to S_3 and choose 7 instead of 5 ; continuing the backtrack process we eventually (after backtracking again to S_4) complete S_7 with the representatives

$$S_1\text{-}1, \; S_2\text{-}3, \; S_3\text{-}7, \; S_4\text{-}5, \; S_5\text{-}8, \; S_6\text{-}6, \; S_7\text{-}2$$

At this stage both the elements 2 and 6 of S_8 have been used and we must start the backtrack procedure again. The final answer for the first system of distinct representatives is

$$S_1\text{-}4, \; S_2\text{-}3, \; S_3\text{-}7, \; S_4\text{-}5, \; S_5\text{-}1, \; S_6\text{-}8, \; S_7\text{-}2, \; S_8\text{-}6$$

i. e. $\{4, 3, 7, 5, 1, 8, 2, 6\}$; the next one is $\{4, 5, 7, 2, 1, 8, 3, 6\}$. These are the only two sets of distinct representatives in this case.

The parameters used in the backtrack algorithm are the same as those used in Hall's algorithm. There are also two local arrays:

An integer array pointer(1::n) so that pointer(i)=j when we are referencing the jth element of the ith set.

A logical array used(1::n) indicates whether an integer is currently in use as a distinct representative. If used(X)=**true** then X is being used as a representative.

When we backtrack care must be taken to reset the pointer for the ith set to the beginning of that set and also to remove the representative of the (i-1)th set.

```
procedure SDR_backtrack (integer value n; integer array S(*, *);
                integer array SDR, members (*); procedure FAIL);
begin integer array pointer(1::n); logical array used(1::n);
        integer i, X;
```

```
        for j:=1 until n do begin pointer(j):=0; used(j):=false
                        end;
        SDR(1):=S(1,1); used(SDR(1)):=true; pointer(1):=1; i:=2;
NEXT: while i≤n do
        begin LOOP: pointer(i):=pointer(i)+1;
                    if pointer(i)≤members(i) then
                      begin X:=S(i,pointer(i));
                            if used(X) then goto LOOP
                            else begin comment X is the new distinct
                                  representative for set i;
                                  SDR(i):=X; used(X):=true; i:=i+1
                                  end
                      end
                    else comment set i has no more elements so we
                          backtrack;
                      if i=1 then FAIL else
                        begin pointer(i):=0; i:=i-1;
                                used(SDR(i)):=false; goto LOOP
                        end
        end;
        comment a new set of distinct representatives is in the array SDR
        we now backtrack for the next set;
        i:=i-1; used(SDR(i)):=false; goto NEXT
end SDR_backtrack;
```

7.4 MATRICES OF ZEROS AND ONES

Matrices of zeros and ones appear frequently in combinatorial problems. In essence they are logical matrices with only two values allowed for the elements, which can therefore be considered as true or false. Konig's basic theorem about such matrices effectively concerns systems of distinct representatives.

Theorem (D. Konig)

If the elements of a rectangular matrix are zeros and ones, the minimum number of lines that contain all of the ones is equal to the maximum number of ones that can be chosen with no two on the same line. (A 'line' of a matrix is either a row or a column.)

Proof. Let $A = (a_{ij})$ be an $n \times t$ matrix of zeros and ones. Let m be the minimum number of lines containing all the ones. Let M be the maximum number of ones that can be chosen so that no two are on the same line.

Obviously $m \geq M$ since no line can pass through two of the ones in M. We will now use Philip Hall's theorem to prove $M \geq m$. Suppose the minimum number of lines m consists of r rows and s columns $(m = r+s)$. Consider these r rows and construct the sets $R_1, R_2, \ldots, R_r$, one for each row. R_i contains the number j as an element if $a_{ij} = 1$ and j is not one of the s columns. If k of the sets $R_1, R_2, \ldots, R_r$ contain between them $v < k$ elements then we could have used these v columns instead of the k rows and thus obtained a smaller m. But we assumed m was a minimum and therefore condition C of Phillip Hall's theorem must hold for the r rows. Hence we can choose r ones from these rows with none from the same columns (and we know by definition that none came from the s columns). By the same argument we may construct sets for the s columns and thus choose s ones, no two on the same row and none from the r rows. The $r+s = m$ ones chosen have the property that none lies on the same line, therefore $M \geq m$. Since we previously showed $m \geq M$, it is established that $M = m$.

7.4.1 Ramsey's Theorem

The pigeon-hole principle in mathematics asserts that if a set of sufficiently many elements is partitioned into not too many subsets then at least one of the subsets must contain many elements.

Ramsey's theorem may be regarded as a profound generalisation of this simple principle.

Suppose we have the graph shown in figure 7.1 below, which consists of six points P_1, P_2, P_3, P_4, P_5 and P_6 each of which is joined to all the others to form a complete graph.

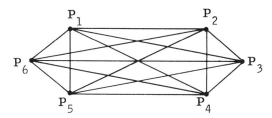

Fig. 7.1. A complete graph with six points

If the lines of the graph are coloured either red or blue we seek to show that there must be three points such that the lines of the triangle they form are all of the same colour. We show this in the following way. Consider any point P_1 (say); there are five lines emanating from it and thus at least three must be of the same colour. Suppose $P_1 P_2$, $P_1 P_3$, $P_1 P_4$ to be coloured red. Therefore in order to prevent a red triangle the lines $P_2 P_3$, $P_3 P_4$ and $P_4 P_2$ must all be blue and thus form a blue triangle.

The generalisation of this problem reveals a result known as Ramsey's theorem, which is stated here without proof.

Theorem (Ramsey)

Let S be a set containing N elements and suppose that the family T of all subsets of S containing exactly r elements is divided into two mutually exclusive families α and β. Let $p \geq r$, $q \geq r$, $r \geq 1$. Then, if $N \geq n(p, q, r)$, a number depending solely on the integers p, q, r and not on the set S, it will be true that there is either a subset A of p elements, all of whose r subsets are in the family α, or there is a subset B of q elements, all of whose r subsets are in the family β.

The proof of this theorem is quite complex and the reader is referred to Marshall Hall's book. An application of Ramsey's theorem to convex bodies, which have the property that any line segment joining two points lies entirely in the body, is as follows: For a given integer n, there is an integer $N = N(n)$ such that any N points in a plane, no three in a line, will contain n points forming a convex n-gon.

7.5   ASSIGNMENT PROBLEMS

The classical assignment problem concerns n men and n jobs and the costs of man i doing job j for all i, j. We are required to assign each man to a different job so that the total cost is a minimum. For example consider three men Brown, Jones and Smith who are required to decorate three houses A, B and C with the cost matrix as follows:

Man	House A	B	C
Brown	5	8	9
Jones	20	18	16
Smith	13	15	14

The minimum cost is 36 with Brown, Jones and Smith decorating A, C and B respectively. One obvious, but inefficient way, to solve such a problem is to consider the n! possible permutations and find the cheapest. However, better ways of solving such problems exist which use the basic results given in the previous sections of this chapter.

7.5.1 The Classical Assignment Problem

The simplest general form of the problem is to take n men and n jobs and an n×n cost matrix C_{ij} (i, j = 1, 2, ..., n) in which c_{ij} is the cost of man i doing job j. We assume at first that all the c_{ij} are positive integers and we wish to assign each man to a different job so as to minimise the sum of the costs. There are n! possible permutations but in order to avoid examining such a large number we adopt a different strategy called the Hungarian method, the basis of which is the following theorem.

Theorem

Given a matrix $C = (c_{ij})$, the minimum of the sum $c_{1j_1} + c_{2j_2} + \ldots + c_{nj_n}$ over all possible permutations $(j_1 j_2 \ldots j_n)$ of the integers 1, 2, ..., n is equal to the maximum of

$$\sum_{i=1}^{n} u_i + \sum_{j=1}^{n} v_j$$

for all numbers u_i (i = 1, 2, ..., n) and v_j (j = 1, 2, ..., n) such that $u_i + v_j \leqslant c_{ij}$ in all cases. This common value is attained when $u_i + v_{j_i} = c_{ij_i}$ (i = 1, 2, ..., n) and these values solve the assignment problem.

Proof. Let

$$m = \text{Max} \left(\sum_{i=1}^{n} u_i + \sum_{j=1}^{n} v_j \right)$$

$$M = \text{Min} \left(\sum_{i=1}^{n} c_{ij_i} \right)$$

Take $u_i = 0$ for all i and $v_j = \text{Min}(c_{ij})$ in each column. Thus $u_i + v_j \leq c_{ij}$ is satisfied.

Therefore $\Sigma u_i + \Sigma v_j \leq \Sigma c_{iji}$, and $m \leq M$.

Form the matrix $B = (b_{ij})$ such that $b_{ij} = C_{ij} - (u_i + v_j)$ for all i and j.

If there are n zeros in B, no two on the same line, these positions give n values of c_{iji} such that $\Sigma u_i + \Sigma v_j = \Sigma c_{iji}$; thus $m = M$ and the problem is solved.

If there are fewer than n zeros in B, no two on the same line, then by Konig's theorem there are $t < n$ lines which include all these zeros. Suppose $t = r+s$ and that these consist of r rows and s columns. Make the following changes:

$$u_i^* = u_i \qquad \text{for the } r \text{ rows}$$
$$u_i^* = u_i + 1 \qquad \text{for the other rows}$$
$$v_j^* = v_j - 1 \qquad \text{for the } s \text{ columns}$$
$$v_j^* = v_j \qquad \text{for the other columns}$$

If we now examine the B matrix there are three cases.

(a) b_{ij} is not on an r row or an s column. $u_i^* + v_j^* = u_i + v_j + 1$. $b_{ij} > 0$ in this case and is at most reduced by 1. Therefore $u_i + v_j \leq c_{ij}$.

(b) b_{ij} is on an r row and an s column. $u_i^* + v_j^* = u_i + v_j - 1$. Since originally $u_i + v_j \leq c_{ij}$ this remains true.

(c) b_{ij} is on an r row and not an s column or is on an s column but not on an r row. In this case $u_i^* + v_j^* = u_i + v_j$, the situation is unchanged and $u_i + v_j \leq c_{ij}$.

So the B matrix still obeys the conditions of the problem that all its elements are ≥ 0. Also

$$\Sigma u_i^* + \Sigma v_j^* = \Sigma u_i + \Sigma v_j + (n-r)-s$$

$$= \Sigma u_i + \Sigma v_j + n-(r+s)$$

Since $n > r+s$ by Konig's theorem

$$\Sigma u_i^* + \Sigma u_j^* > \Sigma u_i + \Sigma v_j$$

So we have increased our total by an integer. We continue the process until we cannot increase $\Sigma u_i + \Sigma v_j$ any more in which case $t = n$ and by Konig's theorem we have a solution for the assignment problem.

Corollary. The theorem is still true if the matrix $C = (c_{ij})$ contains any rational numbers.

We can see that this extension is possible because the solution is not altered if we either add a constant number throughout nor if we multiply each c_{ij} by a constant.

7.5.2 An Algorithm for the Classical Assignment Problem

The algorithm is based on the theorem we have just proved.

Let the cost matrix $C = (c_{ij})$ be such that $c_{ij} \geq 0$ for all $i, j = 1, 2, \ldots, n$. If this is not so originally we can add a suitable number to all the c_{ij}. The algorithm is required to find a minimum cost assignment with each man doing a different job.

Step 1. Find the smallest value in column j and subtract it from all the elements in column j for $j = 1, 2, \ldots, n$. Examine all the rows: if any row does not contain a zero then find the smallest element in that row and subtract it from all the elements in that row. Call this resulting matrix $B = (b_{ij})$.

Step 2. Examine B. If there are n zeros, one in each row and no two in the same column, then the problem is solved and the positions of these n zeros is the minimum assignment. Stop.

Step 3. If there are not n such zeros, by Konig's theorem we find $t < n$ lines passing through all the zeros. Let the t lines be r rows and s columns.

Step 4. Find the smallest element not in an r row or an s column. Let its value be x.

Step 5. Transform the matrix B as follows:

    (i)   If b_{ij} is not on an r row or an s column

$$b_{ij} := b_{ij} - x$$

    (ii) If b_{ij} is on both an r row and an s column

$$b_{ij} := b_{ij} + x$$

(iii) Otherwise b_{ij} is unaltered.

Go to step 2.

The major difficulty in implementing this algorithm efficiently is in steps 2 and 3. In essence it is very similar to finding a set of distinct representatives and if we can find a set these are the positions of the zeros which solve the problem. If we cannot find such a set we must find the largest collection of sets for which distinct representatives exist. Thus we can find the t lines that pass through all the zeros.

Example. Apply the algorithm to the following cost matrix.

			Job		
Man	1	2	3	4	5
1	10	8	7	6	3
2	10	7	7	5	2
3	6	6	4	4	1
4	12	9	7	8	2
5	10	8	5	5	1

Subtracting $6, 6, 4, 4$ and 1 from columns 1 to 5 respectively and then $2, 1$ and 1 from rows $1, 2$ and 4 respectively we obtain the following B matrix:

$$B = \begin{pmatrix} 2 & 0 & 1 & 0 & 0 \\ 3 & 0 & 2 & 0 & 0 \\ 0 & 0 & 0 & 0 & 0 \\ 5 & 2 & 2 & 3 & 0 \\ 4 & 2 & 1 & 1 & 0 \end{pmatrix}$$

There are four lines which pass through all the zeros as indicated:

$$B = \begin{pmatrix} 2 & 0 & 1 & 0 & 0 \\ 3 & 0 & 2 & 0 & 0 \\ 0 & 0 & 0 & 0 & 0 \\ 5 & 2 & 2 & 3 & 0 \\ 4 & 2 & 1 & 1 & 0 \end{pmatrix}$$

The smallest value not in the marked rows is 1. Transform the B matrix as in step 5 of the algorithm:

$$B = \begin{pmatrix} 1 & 0 & 0 & 0* & 0 \\ 2 & 0* & 1 & 0 & 0 \\ 0* & 1 & 0 & 1 & 1 \\ 4 & 2 & 1 & 3 & 0* \\ 3 & 2 & 0* & 1 & 0 \end{pmatrix}$$

Now the zeros marked with an * form $n = 5$ zeros, no two in the same row or the same column. This solves the assignment problem and the minimum value is

$$6 + 7 + 6 + 2 + 5 = 26$$

This solution is not unique and another minimum solution is obtained by allowing man 1 to do job 2 and man 2 to do job 4.

The theorem and algorithm given above for the classical assignment problem found the assignment with minimum cost. When the assignment with maximum cost is required similar methods are used. Two basic approaches are possible.

(a) Form a new cost matrix $C*$ by subtracting each term of C from the largest cost in the original matrix. Now the minimum cost assignment of $C*$ is the maximum cost assignment of C. For example the assignment problem in section 7.5 has the following $C*$ cost matrix

$$\begin{pmatrix} 15 & 12 & 11 \\ 0 & 2 & 4 \\ 7 & 5 & 6 \end{pmatrix}$$

(b) Use the algorithm in section 7.5.2 but start by subtracting the largest value in each column from each element in that column. This produces a B matrix with zeros and negative values. The transformation of the B matrix is now very similar to that given except the largest value (i. e. the negative value nearest to zero) not in an r row or an s column is reduced to zero.

Exercises [7. 7] and [7. 8] can be used to test these two methods.

7.5.3 The Bottleneck Assignment Problem

Consider the following problem: n mechanics have to repair n machines and the time T_{ij} for mechanic i to mend machine j is given. Assign one mechanic to each machine so as to minimise the time until all the machines are in working order.

This problem is slightly different from the classical assignment problem since the criterion now is that the slowest man is as fast as possible - a minimax problem. These problems are called bottleneck assignment problems and are somewhat easier than the classical assignment problems. We can again assume the elements of the matrix to be positive integers and that the integers start with a lowest value and increase in discrete increments; some integers may of course be repeated.

The basic method is to include only the smallest integers and test if there is a possible assignment; if not add in the next smallest integers into the matrix T_{ij} and recheck for an assignment. Continue adding integers until an assignment is found. Consider the example given previously where the cost matrix was

$$\begin{pmatrix} 10 & 8 & 7 & 6 & 3 \\ 10 & 7 & 7 & 5 & 2 \\ 6 & 6 & 4 & 4 & 1 \\ 12 & 9 & 7 & 8 & 2 \\ 10 & 8 & 5 & 5 & 1 \end{pmatrix}$$

We start with a matrix with only the two ones in, which obviously has no solution. We continue adding integers until we reach the following stage:

$$\begin{pmatrix} . & . & 7 & 6* & 3 \\ . & 7* & 7 & 5 & 2 \\ 6* & 6 & 4 & 4 & 1 \\ . & . & 7 & . & 2* \\ . & . & 5* & 5 & 1 \end{pmatrix}$$

An assignment can now be found and the asterisked positions indicate the assignment of the mechanics to the machines. The time

taken before all the machines are mended is 7 . Thus mechanic 2
doing job 2 is the 'bottleneck'.

The most straightforward algorithm for the bottleneck assignment
problem is based on Hall's theorem for finding distinct representatives.
The sets R_i are the n rows $1, 2, \ldots, n$ and the elements of these
sets are the column positions of the T_{ij} . Thus the element j is
included in set i if T_{ij} is included in the matrix at this stage. We
start by including only the smallest element and those equal to it in the
matrix T_{ij} . Then we try to find a set of distinct representatives for
the row sets R_i by using Hall's algorithm. If none exists we add the
next largest elements to the matrix and modify the row sets and again
try to find a set of distinct representatives. We continue adding new
values until a set of distinct representatives is obtained and this is then
the solution of the bottleneck assignment problem.

When we finally add sufficient values to the matrix to obtain a
solution we may find there are several possible solutions. In the
example above the answer was given as mechanics 1 to 5 doing jobs
$(4, 2, 1, 5, 3)$ respectively; but other solutions $(3, 2, 1, 5, 4)$, $(4, 2, 1, 3, 5)$
and $(5, 2, 1, 3, 4)$ also satisfy the bottleneck conditions. In such cases
we can either be satisfied with any solution or select the solution with
minimum cost, i. e. $(4, 2, 1, 5, 3)$.

7. 5. 4 Stable Marriage Assignment

The classical assignment problem uses a cost matrix which
describes the preferences of the men and the relative strengths of those
preferences. However it is not always easy or desirable to measure
such preferences as a number and in some problems it seems more
natural to express them as choice lists. For example, an individual
might list the jobs in the order in which he would prefer to do them.
One of the earliest applications of such an idea was to student admissions
to colleges. In this case both the students and the colleges had choice
lists and the assignment was done under the criterion which the original
authors Gale and Shapley called 'stable marriage'.

Consider two disjoint sets A and B and suppose the members of
A list the members of B in their order of preference and the members
of B do the same. Then an assignment of members of A to the

members of B is said to be a stable marriage if and only if there exists no elements a and b (belonging to A and B respectively) who are not assigned to each other but who both prefer each other to their present partners.

The sets must be disjoint otherwise a stable marriage may not exist and for this reason the sets are often known as the men (A) and the women (B). We shall start by considering n men and n women and each man (woman) giving a choice list for all the women (men).

Let us consider the following simple example with three men a, b and c and three women A, B and C. The choice list for the men giving their orders of preference is

Man a chooses A B C
Man b chooses B A C
Man c chooses A C B

The choice list for the women is

Woman A chooses b a c
Woman B chooses c b a
Woman C chooses a c b

The pairings aA, bB and cC give a stable marriage since only one man c would consider another woman an improvement but woman A prefers man a to man c. The pairings aC, bA and cB also give a stable marriage, and in this example these are the only two stable marriage assignments. However in larger cases there will normally be several stable marriage assignments and we will distinguish three of these, namely:

Male optimal stable solution. This is the stable solution when every man is at least as well off under it as under any other stable solution.

Female optimal stable solution. This is similar except that the women get their best possible choices.

<u>Minimum choice stable solution.</u>   In this stable solution the sum of the choice numbers of the men and the women is a minimum.   This solution may not be unique, but it provides a sort of unselfish optimum solution giving credit to low choice numbers in both sets.

In the example above the solution aA, bB, cC is the male optimal and the sum of the choice numbers is 10 , whilst the solution cC, bA and cB is the female optimal and has the sum of its choice numbers equal to 11 .

We will now prove that there is always a stable set of marriages and the proof, like that for Philip Hall's theorem on distinct representatives, will be a constructive one and hence will yield an algorithm for finding a stable set of marriages.

Theorem

There is always a stable set of marriages.

Proof.   Let each man propose to the first woman in his choice list.   Each woman who receives more than one proposal rejects all except her favourite from those who proposed to her.   So at the end of the first stage several women are holding provisional choices.

At the next stage all the rejected men propose to the next woman in their choice list.   Again the women with more than one proposal reject all but their favourite from the new proposals and the provisional choice from the previous stage.

Again all the rejected men propose to their next choices and this process continues until all the women have a provisional choice and thus have received at least one proposal.   This will eventually happen since no man can propose to the same woman more than once and as long as there is at least one woman who has not been proposed to there will be rejections and new proposals.   When the stage is reached that all the women have had a proposal then we stop and each woman accepts her provisional choice.

The set of marriages so found is stable since if any man John (say) prefers Mary to his wife then he must have proposed to Mary first and she must have rejected him in favour of someone she liked better.

Therefore since Mary prefers her husband to John there is no instability. This completes the proof.

The method used in proving the theorem can easily be adapted into an algorithm for finding a set of stable marriages. The set obtained is the male optimal stable solution; obviously by reversing the roles of the men and women and letting the women propose we can obtain the female optimal stable solution. This was the first method used for stable marriage assignment and is known as the Gale and Shapley algorithm.

An alternative method, known as the McVitie-Wilson algorithm, gives the same male optimal stable solution and goes through the same proposals but does it in a slightly different order which is more efficient. In the McVitie-Wilson algorithm we define the two operations proposal and refusal.

Proposal makes the next proposal for man i and calls the operation refusal for the woman to whom man i has just proposed (proposal does nothing if it is the dummy man zero who is proposing).

Refusal decides for woman j whether the new man i , who has just proposed, is preferable to the man she is holding in suspense. She holds her favourite and for whichever of the two men she rejects she calls the operation proposal so that this man can make his next proposal.

These operations are essentially recursive, and the detailed implementation of them is given below. The procedure MW contains the two operations as procedures PROPOSAL and REFUSAL and uses them by each man in turn proposing to his first choice.

```
procedure MW (integer value n; integer result count;
              integer array malechoice, femalechoice (*, *);
              integer array marriage (*));
```
comment this procedure finds the male optimal stable solution by the McVitie-Wilson algorithm and leaves the result in the array marriage. Thus marriage (i) is the man whom the ith woman marries. There are n men and n women. count is the number of proposals made during

the assignment. malechoice and femalechoice are the choice matrices for the men and women respectively, i.e. malechoice(i, j) is the jth choice of man i. The femalechoice array is changed into the integer array fc where fc(i, j) is the choice number (first, second, third...) of the jth man to woman i. This rearrangement is helpful when a woman is comparing proposals. All women have a dummy man 0 as their initial assignment. This dummy man has a choice number $n+1$ so that he will be given up as soon as any other offer is made;

```
begin procedure PROPOSAL (integer value i);
        comment in this procedure man i makes his next proposal and
        calls procedure REFUSAL. If man i is the dummy man 0 then
        the procedure does nothing;
        if i≠0 then
           begin integer x; count:=count+1;
                x:=malecounter(i); malecounter(i):=x+1;
                REFUSAL(i, malechoice(i, x))
           end;


        procedure REFUSAL (integer value i, j);
        comment this procedure decides whether woman j should keep
        the man she is holding in suspense in marriage(j) or man i who
        has just proposed to her. Whichever she rejects goes back to
        procedure PROPOSAL to make his next proposal;
        if fc(i, marriage(j)) > fc(j, i) then
           begin integer x; x:=marriage(j); marriage(j):=i;
                        PROPOSAL (x)
           end
        else PROPOSAL(i);


        integer array fc(1::n, 0::n);
        integer array malecounter(1::n);
           for i:=1 until n do
           begin for j:=1 until n do
                fc(i, femalechoice(i, j)):=j;
                marriage (i):=0; malecounter(i):=1; fc(i, 0):=n+1
           end;
```

186

```
count:=0;
for i:=1 until n do PROPOSAL(i);
comment this for statement operates the algorithm and after the
ith cycle a set of stable marriages exist for the men 1 to i and
i of the women;
```
end MW;

The Gale and Shapley algorithm is a little less efficient than MW because at each stage it has to examine all the men to see if they have been refused and all the women to see if they have had a new proposal or proposals. In the McVitie-Wilson algorithm each proposal points to the woman being proposed to, and each refusal points to the man being refused so he can make his next choice.

The problem of finding all the stable marriage solutions is a more difficult one and two algorithms have been developed. The McVitie-Wilson algorithm has been extended using the following theorem: In any stable marriage no woman receives a poorer choice than the one she receives in the male optimal stable solution. A simpler, if less efficient algorithm has been given by Wirth who uses essentially a straightforward backtrack approach. His method has the interesting property of obtaining the male optimal solution first and the female optimal solution last (feminists have an obvious modification available).

Unequal Sets and Incomplete Choice Lists

The previous algorithms dealt with the case of n men and n women and complete choice lists. This is not always the case in practice; for example if we apply the method to assigning students to universities then it would be very tedious for each student to include all university courses in his choice list. Incomplete choice lists can be dealt with in several ways; for example

(a) as soon as all the choices in his list have been rejected the person can be removed.

(b) the choice list can be made up either by putting the remaining choices in random order or considering them all equally good.

When the numbers in the two sets A and B are not the same then we can adapt either the Gale and Shapley algorithm or the McVitie-Wilson algorithm to such cases.

Let set A have n members $a_1, a_2, \ldots, a_n$ and set B have m members $b_1, b_2, \ldots, b_m$.

The algorithms are simpler if we let the smaller set propose. Suppose $n \leq m$. The McVitie-Wilson algorithm is:

Step 1. Set $i := 0$.

Step 2. Set $i := i+1$; if $i > n$ then stop.

Step 3. Man a_i proposes to the next woman (b_j say) in his choice list.

Step 4. If woman b_j has had no previous proposal she provisionally accepts a_i ; return to step 2.

Step 5. If woman b_j has a provisional man, a_k , then she chooses between them, keeping the one she prefers as her provisional choice.

Step 6. The rejected man proposes to his next choice (b_j say); go to step 4.

Similarly the Gale and Shapley algorithm can be adapted; the stopping condition is effectively changed so that the algorithm finishes as soon as n women have received proposals.

The algorithm can also be changed so that the larger set proposes but the stopping conditions are more complicated.

There is one very interesting theorem for unequal sets which is not too difficult to prove which states: 'If any person is unmarried in one stable marriage solution he or she will be unmarried in all stable solutions. ' This means that if we apply such algorithms to university admissions no matter what strategy is used (student optimal, university optimal, or something in between) the same students will get places.

7. 6 BIBLIOGRAPHY

The early sections of this chapter on systems of distinct representatives, Konig's theorem, Ramsey's theorem and the classical assignment problem are based on two books by Marshall Hall [1, 2]. The original paper by Philip Hall [3] is brief and lucid. The classical assignment problem has received much attention in recent years mainly through investigators trying to increase the efficiency of the algorithms. One of the earlier algorithms was given by Silver [4].

Stable marriage assignment was originally developed by Gale and

Shapley [5] who examined the problem of student admissions to colleges. Their work was extended by McVitie and Wilson [6, 7] who have given detailed algorithms for equal sets in [8] and for unequal sets in [7]. The algorithms for finding all the stable marriage solutions are in McVitie and Wilson [7, 8] and the one using backtrack in Wirth [9].

[1]   Marshall Hall, Jr. : 'A Survey of Combinatorial Analysis. Chapter 3, Theorems on Choice, pp60-75 in 'Some Aspects of Analysis and Probability. ' John Wiley, 1958.

[2]   Marshall Hall, Jr. : 'Combinatorial Theory. ' Chapters 5, 6 and 7, pp44-65. Blaisdell Pub. Co. , 1967.

[3]   Philip Hall: 'On Representatives of Subsets. ' J. London Math. Soc. , Vol 10 (1935), pp26-30.

[4]   R. Silver: 'An Algorithm for the Assignment Problem. ' Comm. ACM, Vol 3 (1960), pp3-4 (ACM Algorithm 27).

[5]   D. Gale and L. S. Shapley: 'College Admissions and the Stability of Marriage. ' American Math. Monthly, Vol 69 (1962), pp9-15.

[6]   D. G. McVitie and L. B. Wilson: 'The Stable Marriage Problem. ' Comm. ACM, Vol 14, No 7 (July 1971), pp486-90.

[7]   D. G. McVitie and L. B. Wilson: 'Stable Marriage Assignment for Unequal Sets. ' BIT, Vol 10 (1970), pp295-309.

[8]   D. G. McVitie and L. B. Wilson: 'Three Procedures for the Stable Marriage Problem. ' Comm. ACM, Vol 14, No 7 (July 1971), pp491-2 (ACM Algorithm 411).

[9]   N. Wirth: 'Algorithms + Data Structures = Programs. ' pp148-54. Prentice-Hall, Inc. , 1976.

7. 7   EXERCISES

[7. 1] Describe the backtrack algorithm and show how it could be used to find whether the sets $S_1, S_2, \ldots, S_n$ have a system of distinct representatives and if so to display such a system. Without describing Hall's algorithm state when it would be more appropriate than the backtrack algorithm for the problem of finding distinct representatives.

Given the sets $S_1, S_2, \ldots, S_n$ below find a system of distinct representatives

$$S_1 = \{1, 2, 5\}$$
$$S_2 = \{1, 4, 7\}$$
$$S_3 = \{1, 10\}$$
$$S_4 = \{2, 8\}$$
$$S_5 = \{1, 3, 5, 9\}$$
$$S_6 = \{6, 7\}$$
$$S_7 = \{4, 5, 10\}$$
$$S_8 = \{9, 10\}$$
$$S_9 = \{2, 6\}$$
$$S_{10} = \{3, 8\}$$

(Newcastle 1967)

[7.2] State P. Hall's theorem on systems of distinct representatives for the sets $S_1, S_2, \ldots, S_n$. Show how it could be applied to the sets

(i)

$S_1 = \{1, 3, 5, 9\}$	$S_6 = \{6, 9\}$
$S_2 = \{1, 4, 5\}$	$S_7 = \{7, 8\}$
$S_3 = \{1, 2, 7\}$	$S_8 = \{4, 10\}$
$S_4 = \{1, 6\}$	$S_9 = \{2, 5, 6\}$
$S_5 = \{4, 10\}$	$S_{10} = \{3, 8\}$

(ii)

$S_1 = \{B, D, G\}$	$S_5 = \{C, E, H\}$
$S_2 = \{A, E, F\}$	$S_6 = \{B, D, G\}$
$S_3 = \{B, D, E, G\}$	$S_7 = \{D, E, G\}$
$S_4 = \{A, E, F\}$	$S_8 = \{D, G\}$

The sets $S_1, S_2, \ldots, S_n$ contain respectively $2, 3, 4, \ldots, n+1$ elements. Show that there are at least 2^n systems of distinct representatives. Give an example of a system where there is exactly 2^n systems of distinct representatives.

(Newcastle 1969)

[7.3] Let P_1 and P_2 be two partitions of a set of n elements, both of which contain exactly r disjoint subsets. State the necessary and sufficient condition for the possibility of selecting r of the n elements such that the r disjoint subsets in P_1 as well as the r disjoint subsets in P_2 are represented.

190

Let the set of n elements be the numbers $1, 2, \ldots, 20$. P_1 is
the partition of these numbers into four disjoint subsets according to
their remainders when they are divided by 4 . P_2 is the partition into
four disjoint subsets according to the number of prime factors they
contain (e. g. 13 contains no prime factors and 12 contains three).
Is it possible to select four numbers to represent both P_1 and P_2 ?
If so give four such numbers.   Would the same result be true for the
integers $1, 2, \ldots, 16$?

[7. 4] The columns of the following array are the n sets $S_1, S_2, \ldots, S_n$.

$$
\begin{array}{cccccccc}
1 & 2 & 3 & 4 & \ldots & n-2 & n-1 & n \\
2 & 3 & 4 & 5 & \ldots & n-1 & n & 1 \\
3 & 4 & 5 & 6 & \ldots & n & 1 & 2
\end{array}
$$

By considering the cases for some small values of n (e. g. $n = 1, 2, 3, 4$)
and examining the numbers and compositions of the different sets of
distinct representatives, establish a recurrence relation for the number
of such different sets and hence derive the corresponding expression for
that number.

[7. 5] $S_1, S_2, \ldots, S_n$ are n sets which have an SDR .   Suppose that the
elements $a_1, a_2, \ldots, a_r$ $(r < n)$ are an SDR for the sets $S_1, S_2, \ldots, S_r$.
Prove that there is an SDR for the sets $S_1, \ldots, S_n$ which includes the
elements $a_1, a_2, \ldots, a_r$ although not necessarily as representatives of
$S_1, \ldots, S_r$.
    Give an example of sets $S_1, \ldots, S_n$ and elements $a_1, a_2, \ldots, a_r$
which are an SDR for sets $S_1, \ldots, S_r$ but which cannot represent these
sets in any SDR for $S_1, \ldots, S_n$.

[7. 6] In the complete graph with six points shown in figure 7. 1 we colour
the lines either red or blue.   A triangle in the graph is called chromatic
if all its three lines are of the same colour.   Show that there are at least
two chromatic triangles.   Colour the lines in figure 7. 1 so that there are
only two chromatic triangles.
    If the complete graph is extended to seven points what is now the
least number of chromatic triangles?

[7. 7] In the classical assignment problem with five men and five jobs the cost matrix C_{ij} $(i, j = 1, 2, 3, 4, 5)$ contains the twenty-five numbers $1, 2, 3, \ldots, 25$ in some order. We wish to find the assignment with the maximum cost. Show that the maximum is at least 65, and give a cost matrix for which 65 is the optimal value.

[7. 8] Consider the following cost matrix:

				Job			
Man	1	2	3	4	5	6	7
1	10	5	4	6	10	9	3
2	12	8	6	6	12	13	8
3	4	5	5	3	5	5	1
4	4	4	2	3	4	3	2
5	12	10	6	10	10	14	7
6	7	7	10	7	10	7	3
7	5	5	6	6	5	4	4

(a) Find the assignments of the men to the jobs with both maximum and minimum cost. In each case state whether the solution is unique and if it is not give all the possible solutions.

(b) Solve the bottleneck assignment problem using the above cost matrix.

[7. 9] Assuming Hall's theorem on systems of distinct representatives state and prove Konig's theorem for matrices of zeros and ones.

Consider the following assignment problem.

		Job		
Man	1	2	3	4
1	5	7	7	3
2	5	6	5	2
3	7	6	6	3
4	9	8	10	6

The figures in the assignment matrix represent the cost of the man i

doing job j . What assignment of men to jobs would give a minimum cost? What is the cost? Is there more than one minimum?

(Newcastle 1968)

[7.10] State what you understand by assigning the members of two disjoint sets using the principle of stable marriage. Give an algorithm for the stable marriage assignment. What further conditions does your algorithm introduce? Illustrate the working of your algorithm by assigning the 6 students Brown (B), Green (G), Jones (J), Smith (S), Taylor (T) and Wilson (W) to the 5 universities Durham (D), Edinburgh (E), Leeds (L), Newcastle (N), Oxford (O) given the following choice lists:

Brown chooses in the order	E L O N D
Green chooses in the order	O D E L N
Jones chooses in the order	N O L D E
Smith chooses in the order	E O N D L
Taylor chooses in the order	O N D L E
Wilson chooses in the order	O E L D N

Durham chooses in the order	B G T S J W
Edinburgh chooses in the order	W B G T S J
Leeds chooses in the order	G S T J W B
Newcastle chooses in the order	W G S J T B
Oxford chooses in the order	B G T W J S

Assuming each university can take one student, how are the students assigned and which student is not assigned?

(Newcastle 1971)

[7.11] Prove the following two theorems.

(i) If any person is unmarried in one stable marriage solution he or she will be unmarried in all stable solutions.

(ii) In any stable marriage no woman receives a poorer choice than the one she has in the male optimal solution.

[7. 12] The set of men $\{a_1, a_2, \ldots, a_n\}$ has n members and the set of women $\{b_1, b_2, \ldots, b_m\}$ has m members and $n > m$. Give an algorithm for finding the male optimal stable solution.

[7. 13] Consider the three men Tom, Dick and Harry and their three girlfriends Ann, Jane and Mary; they have male choice lists and female choice lists as follows:

Tom	- Ann, Jane, Mary	Ann	- Dick, Tom, Harry
Dick	- Jane, Ann, Mary	Jane	- Harry, Dick, Tom
Harry	- Ann, Mary, Jane	Mary	- Tom, Harry, Dick

What is the male optimal stable solution?

Add a fourth man, Charlie. Can you construct a choice list for him and insert him into the girls lists so that, although he is not assigned one of the girls in any stable solution, the previous male optimal stable solution is no longer valid? What is the new male optimal stable solution?

Notes on the Solutions to Exercises

CHAPTER 2

[2.1] $A(-6)^n + B\,2^n$

[2.2] A. Note that the degenerate form of the equation reduces it to one of first order.

[2.3] $(A+Bn)(-1)^n + (C+Dn)\cos n\pi/3 + (E+Fn)\sin n\pi/3$

[2.4] $A3^n + B(-2)^n - n/6 - 1/36$

[2.5] $A2^n + B3^n + 5^n/6$

[2.6] $(A+Bn)(-2)^n + (n-2/3)/9$

[2.7] $A2^n + Bn2^n + C - 3n^2 2^{n-3} + n^3 2^{n-3}/3$

[2.8] Note three different cases: $m = 4$, $A4^n + B(-1)^n + n4^{n-1}/5$; $m = -1$, P.S. $= (-1)^n n/5$; $m \neq -1, 4$, P.S. $= m^n/(m-4)(m+1)$

[2.9] $(\dfrac{n^3}{6} - \dfrac{n^2}{2} + An + B)(-1)^n + \dfrac{1}{4}(n^{(3)} - 3n^{(2)} + \dfrac{9n}{2} - 3)$

[2.10] $p \neq 0$: $m^n/p^2 + (m^2+p^2)^{n/2}(A\cos n\theta + B\sin n\theta)$, $\theta = \tan^{-1}p/m$
$p = 0$: $(n^2/2m^2 + A+Bn)m^n$

[2.11] $(3^n + 1 + n2^n - 2^{n+1})/2$

[2.12] $(3^{n+1} + 2n3^{n-1} - 2^{n+2} + 1)/4$

[2.13] The last two letters must be the same or different; if the same, the first (k-2) letters must be a permutation of the specified type followed by two like letters different from the final one of the (k-2) - hence $(n-1)P_{k-2}$ such permutations. Similarly for the other case.

$$P_k = (n-1)(P_{k-1}+P_{k-2})$$

$$P_k = A\alpha^k + B\beta^k, \quad A = (n^2-n\beta)/\alpha(\alpha-\beta), \quad B = (n^2-n\alpha)/\beta(\beta-\alpha)$$

where α, β are the roots of $x^2 - (n-1)(x+1) = 0$.

[2.14] (a) $n^2/4 - n/2 + A + B(-1)^n$; (b) Find the number of triplets which include the largest integer to obtain relations between (U_{2n+1}, U_{2n}) and (U_{2n}, U_{2n-1}). Combining these relations gives (A) with the appropriate solution (a) with $A = -B = 1/8$.

[2.15] (i) $u_n = A\alpha^n + B\beta^n$, $v_n = 2[(1+\sqrt{2})A\alpha^n + (1-\sqrt{2})B\beta^n]$, $\alpha = 2+\sqrt{2}$, $\beta = 2-\sqrt{2}$

(ii) A valid (n+1)-permutation with n+1, n adjacent, can be obtained from each of the (u_n+v_n) n-permutations by inserting $(n+1)$ on either side of n. Hence $u_{n+1} = 2(u_n+v_n)$. If $(n+1)$ is not next to n, it must be next to $(n-1)$ and so n+1 can go only on one side of $(n-1)$ in u_n cases, but on either side in v_n cases; $v_{n+1} = u_n + 2v_n$, $u_2 = 2$, $v_2 = 0$; $A = (2-\sqrt{2})/2\alpha^2$, $B = (2+\sqrt{2})2\beta^2$.

[2.16] n+1 must be in either the (n+1)th place or the nth; if the former the first n positions form an appropriate permutation (u_n), if the latter the first (n-1) positions do (u_{n-1}).
$u_{n+1} = u_n + u_{n-1}$; $u_1 = 1$, $u_2 = 2$
Fibonacci numbers.

[2.17] See section 2.5.

[2.18] (i) Eliminate u_{2n-1}, u_{2n-3} from three successive equations for $u_{2n}, u_{2n-1}, u_{2n-2}$.
$$a^2u_{2n} + (2ac-b^2)u_{2n-2} + c^2u_{2n-4} = 0$$

(ii) $u_n = A\alpha^n + B\beta^n$; $v_n = [A(\sqrt{5}-1)\alpha^n - B(\sqrt{5}+1)\beta^n]/2$,

$\alpha, \beta = (3 \pm \sqrt{5})/2$; $A = 1/\sqrt{5}$, $B = -1/\sqrt{5}$

(iii) From the elimination in (ii), both u_n, v_n satisfy

$x_{n+2} - 3x_{n+1} + x_n = 0$ and from (i) these are alternate terms

of $f_{n+2} - f_{n+1} - f_n = 0$.

(iv) The first cell is allocated or it is not; if allocated it is

either a block itself and can be followed by $(u_{n-1} + v_{n-1})$

different allocations, or forms part of a block with the first of

the succeeding (n-1) cells, the first of which must be allocated

(u_{n-1}) . Hence $u_n = 2u_{n-1} + v_{n-1}$.

If the first is not allocated, any allocation of the remainder is

valid $(u_{n-1} + v_{n-1})$.

[2.19] $x_n = Am_1^n + Bm_2^n$ where $m_1 = 1+\sqrt{3}$, $m_2 = 1-\sqrt{3}$. The boundary

conditions show $A \neq 0$ (there is no need to find its value!) and

hence $x_{n+1}/x_n \to Am_1^{n+1}/Am_1^n = m_1$.

[2.20] Solve the difference equations, showing that $v_n = A\alpha^n + B\beta^n$,

$u_n = v_{n+1} - v_n$ where α, β satisfy $x^2 - 2x - 1 = 0$ so that

$u_n = \sqrt{2}(A\alpha^n - B\beta^n)$ with $\alpha = 1+\sqrt{2}$, the larger root. Hence

$u_n/v_n \to \sqrt{2}$. Substitution of numerical values in u_n/v_n show

that 14 iterations are needed to obtain 10 decimal places.

[2.21] Such binary sequences must end in either zero or one $(a_n, b_n$ in

number respectively). Hence $a_{n+1} = b_n$, $b_{n+1} = a_n + b_n$ and

b_{n+1} thus gives the total number of the sequences.

[2.22] n-digit sequences fall into four classes: those containing an

even (odd) number of zeros and an even (odd) number of ones

(a_n, b_n, c_n, d_n) . Show that

$$a_{n+1} = 2a_n + b_n + c_n , \quad b_{n+1} = a_n + 2b_n + d_n ,$$

$$c_{n+1} = a_n + 2c_n + d_n , \quad d_{n+1} = b_n + c_n + 2d_n ,$$

and hence that $a_n = 2^{n-1} + 4^{n-1}$.

197

[2.23] n-sequences with no occurrence of 010 end in 00, 01, 10 or 11 (a_n, b_n, c_n, d_n). The number of sequences we require is then b_{n-1}. Show that

$$a_{n+1} = a_n + c_n, \quad b_{n+1} = a_n + c_n, \quad c_{n+1} = d_n, \quad d_{n+1} = b_n + d_n.$$

Solve, giving $b_n = A\alpha_1^n + B\alpha_2^n + C\alpha_3^n$ where $\alpha_1, \alpha_2, \alpha_3$ are roots of $x^3 - 2x^2 + x - 1 = 0$ $(\alpha \simeq 1.7549, \ \alpha_2, \alpha_3 \simeq 0.1226 \pm 0.7449i)$.

[2.24]
$$\begin{pmatrix} 3^{100} & 2^{100} - 3^{100} \\ 0 & 2^{100} \end{pmatrix}; \ \text{use} \ \begin{pmatrix} a_{n+1} & b_{n+1} \\ c_{n+1} & d_{n+1} \end{pmatrix} = \begin{pmatrix} 3 & -1 \\ 0 & 2 \end{pmatrix} \begin{pmatrix} a_n & b_n \\ c_n & d_n \end{pmatrix}$$

[2.25] $a_n = (\alpha^n - \beta^n)/\sqrt{5}$ where $\alpha, \beta = (1 \pm \sqrt{5})/2$. Use $\alpha\beta = -1$ to show

$$a_n^2 + a_{n+1}^2 = (\alpha - \beta)(\alpha^{2n+1} - \beta^{2n+1})/5 = a_{2n+1},$$

working with α, β's until the final step.

[2.26] (a) $u_n = n \cdot 2^{n-1}$, $G(x) = 2x/(1-2x)^2$.

(b) (i) Note that $(2^m + r)$ has the binary representation of r $(r < 2^m)$ with a one at the most significant end.

(ii) The even members end in zero and the next member has all the same digits apart from the last which is a one.

(iii) $G(2n+1) = 2G(n) + n + 1$.

(iv) $\pi(x) = 2(1+x)\pi(x^2) + x(1-x)^{-2}(1+x)^{-1}$. Difficult to solve since different arguments (x and x^2) appear for the unknown generating function π.

(c) If at the final stage the last two sets have i and $n-i$ items where $i < n-i$ (i.e. $i \le n/2$) the cost of the last merge is i and the previous maximum costs in forming the two sets are $H(i)$ and $H(n-i)$.

CHAPTER 3

[3.1] $A e^{n(n-1)}$

[3.2] $n! + (n-1)!$

[3.3]   The equation is $[(2n-1)\Delta^2 - 4n\Delta + 4]u_n = 0$; if $u_n = 3^n f_n$, f_n satisfies $(6n-3)\Delta^2 f + (4n-4)\Delta f = 0$, and $3(2n-1)v_{n+1} - (2n+1)v_n = 0$ for $v_n = \Delta f_n$.
$v_{n+1} = (2n+1)C/3^{n+1}$, giving $f_n = An/3^n + B$, $u_n = An + B3^n$.

[3.4]   One solution $u_n = A$; if $u_n = Af_n$ $(n+1)\Delta^2 f_n + (n+2)\Delta f_n = 0$ giving

$$f_n = c + c' \sum_{1}^{n-1} (-1)^r/r!$$

[3.5]   By trial one solution is $u_n = (-1)^n/n!$; if $u_n = (-1)^n f_n/n!$, the equation reduces to $\Delta^2 f_n + (n+2)\Delta f_n = 0$, or $v_n = \Delta f_n$ and $v_{n+1} + (n+1)v_n = 0$.

$$u_n = [c + c' \sum_{r=1}^{n} (-1)^r (r-1)!](-1)^n/n!$$

Another approach is to write the equation as $(n+2)u_{n+2} + u_{n+1} = (n+1)u_{n+1} + u_n$ (a form suggested by the trial solution) so that $nu_n + u_{n-1} = A$, a first order equation.

[3.6]   Although a solution $n!$ can be found by inspection, it is easier to write the equation as $u_n - nu_{n-1} = -(u_{n-1} - (n-1)u_{n-2})$ i.e. $v_n = -v_{n-1}$. Hence $u_n - nu_{n-1} = v_n = (-1)^n A$. Solution of this first order equation gives

$$u_n = cn! + c'n! \sum_{0}^{n} (-1)^r/r!$$

[3.7]   The first pair divides the line into two independent pieces and the left hand end of the whole line will be vacant only if the left hand end of the left hand part line is vacant. Hence

$$p_n = [p_0 + p_1 + \ldots + p_{n-2}]/(n-1) \ .$$

The equation of exercise [3.4] follows. Since $p_1 = 1$, $p_2 = 0$, the solution is

$$p_n = \sum_0^n (-1)^r /r! \ ,$$

which tends to e^{-1} rapidly as n increases.

[3.8] Either the left hand site is occupied or it is vacant; if occupied the number of patterns is u_{n-2}, if vacant u_{n-3} since the next pair must be filled. Hence $u_n = u_{n-2} + u_{n-3}$, and $u_1 = u_2 = 1$, $u_3 = 2$. For triplets $u_n = u_{n-3} + u_{n-4} + u_{n-5}$.

[3.9] Separation of the variables leads to $u_{m,n} = m! \, n!$

[3.10] Let u_n = number of ways of arranging the letters so that none is in its correct envelope. The probability is $p_n = u_n/n!$ If another letter, L_{n+1}, and envelope, E_{n+1}, are added either
    (i) L_{n+1} is put in E_r $(r \leq n)$ and L_r is in E_{n+1}
or (ii) L_{n+1} is put in E_r $(r \leq n)$ and L_r is not in E_{n+1}
For (i) we can pick r in n ways and then have u_{n-1} ways of arranging the other $(n-1)$ L's and E's; and for (ii) we can arrange the n E's and L's in u_n ways and then pick an E (in n ways) and exchange its contents with (L_{n+1}, E_{n+1}). Hence $u_{n+1} = nu_n + nu_{n-1}$. The solution is

$$u_n = n! [1 - \frac{1}{1!} + \frac{1}{2!} - \ldots \frac{(-1)^n}{n!}]$$

For r correct, select these in $\binom{n}{r}$ ways, and arrange the others all incorrectly in u_{n-r} ways. Hence

    Prob(r correct) $= p_{n-r}/r!$

i.e. approximately a Poisson distribution, mean 1.

[3.11] Same argument as exercise [3.10], gives u_n .

[3.12] First order equation: $u_n = 2(n+1)H_n - 3n$. Since

$$H_n \sim \int_1^n dx/x = \log n, \quad u_n \sim 2(n+1)\log n .$$

CHAPTER 4

[4.1] 30

[4.2] $5!/2=60$; $4!=24$

[4.3] $32!/(12!)^2 8!$

[4.4] 1092; 728 (excluding zero)

[4.5] Each digit occurs six times in every position: hence
sum $= 60 \times 1111 = 66660$

[4.6] $n(n-1)/2$; $(n^2+n+2)/2$ which is the appropriate solution of
$u_n = u_{n-1}+n$

[4.7] Consider the integers to be of three types $3i$, $3i+1$, $3i+2$. For
the result to be divisible by three we must either have one of each
type $(=n^3)$ or all three of same type $(=3\binom{n}{3})$.
Total $= (3n^3-3n^2+2n)/2$.

[4.8] The five messengers are partitioned in two ways

$$311 \text{ giving } \frac{5!}{3!} \times 3 = 60 \text{ or } 221 \text{ giving } \frac{5!}{2!\,2!} \times 3 = 90$$

Total therefore is 150 .

[4.9] $5!\binom{8}{3} = 6720$

[4.10] The symbols give $6!$ The blanks $\binom{6}{4}$. Total $= 6!\binom{6}{4} = 10800$

[4.11] $4 \times 3^5 = 972$

[4.12] Multiply the G. F. by z and z^2 and subtract these from the original G. F. to obtain

$$(1-z-z^2)G(z) = F_0 + (F_1-F_0)z + (F_2-F_1-F_0)z^2 + \ldots = z$$

[4.13] Add one to each part and the problem becomes one of finding the compositions of $n+m$ with exactly m parts

i. e. $\binom{n+m-1}{m-1}$

[4.14] The generating functions are

$$G_1(x) = (1+x+x^2)(1+x^2+x^4) \ldots (1+x^i+x^{2i}) \ldots$$
$$G_2(x) = (1-x)^{-1}(1-x^2)^{-1}(1-x^4)^{-1}(1-x^5)^{-1}(1-x^7)^{-1} \ldots$$

The result follows from the identity

$$(1+x^i+x^{2i})(1-x^i) = (1-x^{3i})$$

[4.15] Take the Ferrers graph for a partition of n and add a column of n dots on the left.

[4.16] This is equivalent to the compositions of $30-8 = 22$ into eight parts. $\binom{21}{7} = 116, 280$.

[4.17] Use the G. F. 's given in equations (4. 30) and (4. 31).

[4.18] $\binom{6}{3} = 20$ combinations of w, x, y, z

(a) $\binom{6}{4} = 15$, (b) $\binom{9}{4} = 126$, (c) 126 - number with 4 flavours (15) - number with 1 flavour (6) = 105 , (d) (20 ways of picking 3 flavours from 6) $\times$ (3 ways of filling 4 cones with exactly 3 flavours) = 60 .

[4.19] $p_1+p_2 = 2n_1+n_2+n_3 > p_3$ etc. $p_1 > p_2 > p_3 > 0$, and construction is reversible (one-to-one).

[4.20] $\binom{15}{5} \times \binom{12}{2}$; $\binom{10}{5}\binom{6}{2} + \binom{10}{5}\binom{7}{2} + \binom{9}{5}\binom{7}{2} + \binom{9}{5}\binom{6}{2}$; $\binom{11}{7}$.

Since groups are determined once seven integers are selected.

[4.21] Number of integers divisible by p_1 is $(n/p_1 - 1)$, by $p_1 p_2$ is $(n/p_1 p_2) - 1$.

[4.22] (i) $N = n^r$ is the number of ways of distributing r distinct objects into n distinct cells, subtract those with one cell empty $\binom{n}{1}(n-1)^r$, add those with two cells empty $\binom{n}{2}(n-2)^r$, etc...

(ii) The exponential G.F. is $(x + \frac{x^2}{2!} + \frac{x^3}{3!} + \dots)^n = (e^x - 1)^n$

(iii) Substitute $\sum_{i=0}^{n} (-1)^i \binom{n}{i}(n-i)^r$ into the right hand side of the recurrence relation and after a little manipulation the left hand side is obtained.

[4.23] (ii) Let a_i be the property that $i+1$ follows i. We can show that for example $N(a_1) = (n-1)!$ $N(a_1 a_2) = (n-2)!$ etc. and there are $(n-1)$ properties a_i. Using the principle of inclusion and exclusion we arrive (after some manipulation) at the result given.

[4.24] See section 4.4, the proportion is $1/e$.

[4.25] Proof by G.F.'s is possible but the easiest way is to consider the Ferrer graph of a partition of n with exactly k parts and subtract the end dot from each row.

Partitions of 11 with 3 parts   9 1 1, 8 2 1, 7 3 1, 7 2 2, 6 4 1,
                                        6 3 2, 5 5 1, 5 4 2, 5 3 3, 4 4 3

Partitions of 8 no part > 3 1^8, $2\,1^6$, $2\,2\,1^4$, $3\,1^5$, $2^3\,1^2$,
                                     $3\,2\,1^3$, 2^4, $3\,2\,2\,1$, $3^2\,1^2$, $3\,3\,2$

[4.26] 10, 8 2, 7 3, 6 4, $6\,2^2$, 5^2, 5 3 2, $4^2\,2$, $4\,3^2$, $4\,2^3$,
$3^2\,2^2$, 2^5;

     G.F.'s $p(t) = (1-t)^{-1}(1-t^2)^{-1} \dots (1-t^k)^{-1} \dots$
                $p^*(t) = (1-t^2)^{-1} \dots (1-t^k)^{-1} \dots$

Hence $(1-t)p(t) = p^*(t)$ and coefficients of t^n give the result;

or by Ferrer's graph $p_n = p_{n-1} + p_n^*$

Similarly $p_n^{(k)} = p_n - p_{n-k}$

$p_{n+2} - 2p_{n+1} + p_n = p_{n+2}^* - p_{n+1}^* \geq 0$.

[4.28] Two. (AC and BD)

[4.30] A tree.

CHAPTER 5

[5.1] (a) 4166, 746, 1055

(b) $(0,0,0,0,1,4)$ $(0,2,0,4,0,0)$, $(1,2,1,1,4,1)$

[5.2] (a) 2 0 1 4 3 , 4 0 2 3 1

(b) 1 0 3 4 2 , 3 1 2 4 0

(c) 2 0 3 1 4 , 3 1 2 0 4

(d) Factorial representations are $(1,0,2,0)$ and $(0,1,0,0)$
giving permutations 4 1 3 0 2 , 4 2 0 1 3 .

[5.3] An algorithm for reverse lexicographical order similar to the
Mok-kong Shen algorithm is as follows:

Given the permutation $k_1 \, k_2 \, \ldots \, k_n$ the next one is found by the
following:

(1) Find the smallest i such that $k_i < k_{i+1}$.

(2) Find the smallest j such that $k_j < k_{i+1}$.

(3) Interchange k_{i+1} and k_j .

(4) Reverse the order of the digits $k_1 \, k_2 \, \ldots \, k_i$.

For example if we consider 4 2 1 3 then $i = 3$, $j = 2$;
interchanging k_4 and k_2 gives 4 3 1 2 , reversing the digits
$k_1 \, k_2 \, k_3$ gives 1 3 4 2 which is the next permutation.

[5.5] 123 124 132 134 142 143   213 214 231 234 241 243
312 314 321 324 341 342   412 413 421 423 431 432

A possible algorithm for finding the next lexicographical
r-permutation given the current one in $a(1), a(2), \ldots, a(r)$ is

<u>Step 1</u> . Set i:=r+1;

<u>Step 2</u> . i:=i-1; stop if i=0

<u>Step 3</u> . X:=a(i);

<u>Step 4</u> . X:=X+1;

<u>Step 5</u> . if X > n then **goto step** 2

<u>Step 6</u> . if X=a(j) for any j≠i (j=1, 2, ... , r)

then **goto step** 4

else use the new permutation

a(1), a(2), ... , a(i-1), X, and the values in a(i+1), ... , a(r)

are the smallest of the n values not already in the

permutation;

[5.6] The binary sequences are in increasing numerical order. We can therefore successively add one to the previous binary sequence. This will not always produce a valid sequence because it need not have exactly r ones, but if not we continue to add one until a new valid sequence is reached. Stop when the final bit sequence r ones followed by n-1 zeros is obtained; in the example this is the bit sequence (111000) .

[5.7] The method suggested is given in Lehmer [2, p20] and a possible algorithm is

(1) Find the factorial digits $(a_1, a_2, ... , a_{n-1})$ of the integer m .

(2) Use these digits as the column headings of an array as follows

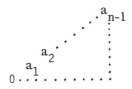

(3) The values below these digits are formed column by column starting at the left. Each column is formed from its left hand neighbour by attaching its new column heading, copying those elements which are less than this heading, and increasing all other elements by one.

(4) The last column is the final permutation.

Example: In exercise [5.1] we showed that 1235 is represented by the factorial digits $(1,2,1,1,4,1)$. We therefore get the array

```
                1
              4 5
          1   1 2
        1 2   2 3
      2 3 4   5 6
    1 1 2 3   3 4
  0 0 0 0 0 0 0
```

The 1235th permutation is therefore 1 5 2 3 6 4 0 .

[5.8] The algorithm for finding next lexicographical r-combination of n objects with unlimited repetition is (given $k_1 \; k_2 \; \ldots \; k_r$ as present r-combination)

(a) Find largest i such that $k_i < n$

(b) $k_i := k_i + 1$;

(c) **for** j:=i+1 **until** r **do** $k_j := k_i$

[5.9] Since there is a one-to-one correspondence between the integers from 0 to $2^{n-1} - 1$ and the compositions, if n is not too large we can generate a random integer in the range 0 to $(2^{n-1} - 1)$ and find the appropriate composition.

[5.12] The recurrence relation is $u_r = (n-1)u_{r-1}$ giving an answer of $u_r = n(n-1)^{r-1}$ since $u_1 = n$.

CHAPTER 6

[6.1] If we take the elements in decreasing order $(5,4,3,2,1)$, backtrack produces the components in the order 5, 4 1, 3 2, 31^2, $\underline{2\,3}$, $2^2 1$, $\underline{2\,1\,2}$, 21^3, $\underline{1\,4}$, $\underline{1\,3\,1}$, $1\,2^2$, $1\,2\,1^2$, $1^2\,3$, $1^2\,2\,1$, $1^3\,2$, $1^3\,2$, 1^5 ; the corresponding tree is:

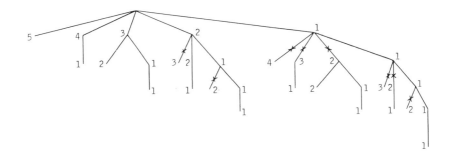

Notice this example of a tree with leaves at different levels - in
fact having the number of leaves at the different levels given by
the binomial coefficients $(1, 4, 6, 4, 1)$ - and nodes with different
numbers of branches emerging from them.

The partitions of 5 are those compositions above which are <u>not</u>
underlined and the tree is as shown with the branches (and any
subtree below) marked with a cross $(\times)$ removed.

[6.2]   1 2 3 4 5, 1 2 3 5 4, 1 2 4 3 5, 1 3 2 4 5, 1 3 2 5 4, 2 1 3 4 5,
        2 1 3 5 4, 2 1 4 3 5.

[6.3]   The 20th is 1 5 4 3 2 ; the permutations excluded are 1 3 4 2 5,
        1 4 3 2 5, 1 5 2 3 4, 1 5 2 4 3.

[6.5] $(1, 4, 3, 2)$ or $(4, 1, 3, 2)$: Cost 22 .

[6.6]   The algorithm should check that each of $1, 2, \ldots, n$ is represented
        r times and with the correct spacing, and should avoid unnecessary
        repeated checking; devices like reversing the sign of integers
        checked and keeping a tally adding 2^{i-1} after the digit i has been
        checked may be helpful.

        In order to estimate what size of problem can be tackled we need
        to estimate how many sequences might need examination. The
        integer n can be put in $(n-1)$ positions, and $n-1$ in the same
        number; subsequently the number of possible positions grows and

then falls again but a crude estimate of the total number might be of the order of $(n-1)^{n-1}$, indicating that $n = 10$ is a big problem.

[6. 7] Preferable to keep in store all the Q_n that will be needed; since $n^3 + 5n \leq 600\,000$ we need about $n = 84$. A backtrack program would need little store but there would be much repetitive testing. A good method adapts the Eratosthenes sieve method for finding primes and builds up the minimum number of Q_n needed for each integer e. g.

Integer	1	2	3	4	5	6	7	8	9	10	11	12	13	14
Least number of Q_n	1	2	1	2		2	1	2		2				1

Thus at the third pass it is found that 5 can be formed from 4 (which needs $2\,Q_n$) and 1 (needs 1), hence $3\,Q_n$. This method would need $100\,000$ words which could be reduced if information were packed.

[6. 8] Minimum bandwidth is 2 .

[6. 9] Note that one of each pair of valid partitions must contain 1 , that the largest part of this partition must exceed the sum of the other parts and that the second partition follows uniquely. Lexicographical enumeration yields $\{(1\ 2\ 4\ 10),(8\ 9)\}$, $\{(1\ 2\ 5\ 9),(4\ 13)\}$ and $\{(1\ 16),(2\ 3\ 4\ 8)\}$.

[6. 10] Best case is where the branches from only one node at each level are explored i. e. $n + (n-1) + (n-2) + \ldots + 1 = n(n+1)/2$. It is conceivable that every node might have to be retained except the n nodes which have been explored

i. e. $\{n + n(n-1) + n(n-1)(n-2) + \ldots + n!\} - n \sim n!\,e$

[6. 12] $2^n - 1$ (each object may be included or excluded, hence 2^n sets but one is empty). Several alternative orderings are possible, but convenient to use one-to-one correspondence with binary

208

integers $1, 2, \ldots, 2^3 - 1$ so that first binary place represents $+$, the second $\times$ etc. Thus $5 = (101)_2 \longleftrightarrow (+ -)$. A backtrack algorithm can be used.

[6.13] Cf. exercise [6.7].

[6.16] 57

[6.17] The minimum order of multiplication is $(M_1 \times (M_2 \times (M_3 \times M_4))) \times M_5$ and this results in 1440 operations.

[6.18] In the method of section 6.3.4 we need to store all $f(i; j_1, \ldots, j_k)$ for $i = j_1, \ldots, j_k$, hence $n \binom{n-1}{k}$ values, which is greatest for $k \sim (n-1)/2$. Thus we need about $n. (n-1)! / \{[(n-1)/2]! \}^2$ locations assuming that we have been able to overwrite almost all of the previous values (not trivial!). Using Stirling's approximation $n! \sim \sqrt{(2\pi)} n^{n+\frac{1}{2}} e^{-n}$, for $n = 2m+1$ we need about $(2m+1) 2^{2m} / \sqrt{(\pi m)}$ stores. For $2m = 10$, i.e. 11 towns, this is ~ 3300.

CHAPTER 7

[7.1] Several solutions e.g. $(1, 7, 10, 8, 5, 6, 4, 9, 2, 3)$ or $(2, 1, 10, 8, 5, 7, 4, 9, 6, 3)$ or $(1, 4, 10, 2, 3, 7, 5, 9, 6, 8)$

[7.2] (i) Solution depends on depth-first or breadth-first search in Hall's algorithm.

| Depth-first | 3, 1, 2, 6, 10, 9, 7, 4, 5, 8 |
| Breadth-first | 3, 5, 1, 6, 10, 9, 7, 4, 2, 8 |

(There could be other slight variations in the Hall algorithm.)
(ii) No SDR. Sets S_1, S_3, S_6, S_7, and S_8 contain only the four elements B, D, E, G.
Proof by induction since each SDR for the $(n-1)$ sets contains $(n-1)$ elements and there are $(n+1)$ elements in set S_n.
Example: $S_1 = (1, 2)$, $S_2 = (1, 2, 3), \ldots, S_n = (1, 2, 3, \ldots, n+1)$

[7.3] The partitions P_1 and P_2 should have a common SDR, and the condition could be stated as follows: no $t < r$ of the disjoint subsets of one partition have all their elements in less than t subsets of the other partition.

Partitions are

P_1 = {1 5 9 13 17}, {2 6 10 14 18}, {3 7 11 15 19 }, {4 8 12 16 20}

P_2 = {1 2 3 5 7 11 13 17 19}, {4 6 9 10 14 15}, {8 12 18 20}, {16}

Answer must include 16 and 18 and then either 9 with (3 or 7 or 11 or 19) or 15 with (1 or 5 or 13 or 17).

There is no set of representations satisfying the conditions for the integers 1, 2, ..., 16.

[7.4] If the number of sets of distinct representatives in the general case is S_n we can show that

$$S_n = 2S_{n-1} - S_{n-3}$$

Using difference equation methods (chapter 2) we can solve the above equation for the boundary conditions $S_1 = 3$, $S_2 = 5$, $S_3 = 6$ to give

$$S_n = 2 + \alpha^n + \bar{\alpha}^n$$

where $\alpha = (1+\sqrt{5})/2$ and $\bar{\alpha} = (1-\sqrt{5})/2$.

Alternative: Note two special cases 1 2 3 4 ... n and 3 4 5 ... n 1 2, ignoring these, all other cases C_n can be expressed as

$$C_n = C_{n-1} + C_{n-2}$$

The Fibonacci series with starting values $C_1 = 1$, $C_2 = 3$ giving $C_n = \alpha^n + \bar{\alpha}^n$.

[7.5] Consider using Hall's algorithm from the stage when $a_1 a_2 \ldots a_r$ represent $S_1, S_2, \ldots, S_r$ respectively and we continue the algorithm with sets S_{r+1} etc. We must eventually arrive at an SDR because one exists and that SDR must include $a_1, a_2, \ldots, a_r$ because if they do not represent $S_1, S_2, \ldots, S_r$ they will have been swapped so that they represent some other set.

Example: $r = 3$, $n = 6$ $S_1, \ldots, S_r$ are $(1, 4)$, $(2, 5)$, $(3, 6)$ represented by $1, 2, 3$. $S_{r+1}, \ldots, S_n$ are $(1), (2), (3)$; the final SDR is therefore $4, 5, 6, 1, 2, 3$.

[7.6] Colour the lines as follows:

red $- P_1P_2, P_1P_3, P_1P_4, P_2P_5, P_3P_5, P_3P_6, P_4P_6, P_5P_6$

blue $- P_1P_5, P_1P_6, P_2P_3, P_2P_4, P_2P_6, P_3P_4, P_4P_5$

There are only two chromatic triangles: red $P_3P_5P_6$, blue $P_2P_3P_4$.

The least number of chromatic triangles for seven points is 3 .

[7.7]
$$C = \begin{pmatrix} 1 & 2 & 3 & 4 & 5 \\ 6 & 7 & 8 & 9 & 10 \\ 11 & 12 & 13 & 14 & 15 \\ 16 & 17 & 18 & 19 & 20 \\ 21 & 22 & 23 & 24 & 25 \end{pmatrix}$$

[7.8] (a) Minimum $= 32$. Two solutions with men assigned to jobs $(2, 4, 1, 5, 3, 7, 6)$ and $(2, 4, 1, 6, 3, 7, 5)$.

Maximum $= 59$. Two solutions $(1, 5, 2, 7, 6, 3, 4)$ and $(5, 1, 2, 7, 6, 3, 4)$

(b) Using minimum values; all numbers up to and including 6 used. Solution $(2, 4, 1, 5, 3, 7, 6)$ and $(2, 4, 1, 6, 3, 7, 5)$ each giving value of 32 , same as classical assignment problem solution.

[7.9] Two solutions with men assigned jobs $(1, 3, 4, 2)$ and $(1, 4, 3, 2)$ with minimum value 21 .

[7.10] Only stable assignment is: Durham - Green, Edinburgh - Wilson, Leeds - Taylor, Newcastle - Smith, Oxford - Brown. Jones not assigned.

[7.11] (i) Let there be n men and k women. Consider what happens when the larger set propose and we obtain the stable marriage solution by either the Gale and Shapley or McVitie-Wilson algorithm. In particular consider the position of the $|k-n|$ members of the larger set who are unmarried in this stable solution.

(ii) Consider what would happen if some woman j received a poorer choice than the man (k) she received in the male optimal solution. Show that both woman (j) and man (k) would prefer each other to their new partners and therefore the new assignment is unstable.

[7.12] This algorithm is the same as the equal size set one except for the stopping condition which is:
The algorithm is stopped when each man is either being kept in suspense by some woman or has been refused by all of the women.

[7.13] Original male optimal stable solution

Tom - Ann, Dick - Jane, Harry - Mary

If Charlie has a choice list - Jane, Ann, Mary, and the girls' lists become

Ann - Dick, Tom, Harry, Charlie
Jane - Harry, Charlie, Dick, Tom
Mary - Tom, Harry, Dick, Charlie

the original solution is unstable. Jane prefers Charlie to Dick

and Charlie prefers Jane to nothing. New male optimal stable
solution

Tom - Mary, Dick - Ann, Harry - Jane

So although Charlie is unmarried he can have a nuisance value!

Index

Address calculation 100
Adjacency matrix 80-1
Adjacency structure 81-2
Adjacent mark order 106-7, 109-11
Aho, A. V. 156-7
Algol expression 26-8, 158
Alphabetical order 97
Assignment 164
   bottleneck 181-2, 192
   classical 158, 164, 175-8,
      188, 192-3, 211
   stable marriage 164, 182-8,
      212-13

Backtrack 131-8, 156-61,
   171-3, 189, 206-9
Baumert, L. D. 156-7, 161
Bellman, R. E. 126, 156-7
Berge, C. 89
Binary sequences (also Bit
   patterns) 34, 118, 128, 197, 205
   representations of combina-
      tions 57-9
Binary sequence search tree 53
Binomial coefficients 40, 57, 61-3
Branching 140-2
Branch and bound 139-49, 156-7,
   159-60

Chain 75
Chromatic triangle 175, 191, 211
Circuit in a graph 75
Clowes, J. S. 52
Coffman, E. G. Jr. 157
Combinations 5, 38, 54-8, 60-2,
   88, 92, 131-2, 202-3
   dual 57
   generation 116-17, 128-9,
      205-6
   reverse 57
Comma free code 156, 161
Complement of a graph 76

Complementary function 14, 16,
   18, 20, 36-7
Complexity 3-4, 154-6
Compositions 67-70, 91, 132
   generation 118-20, 126, 157
Connected graph 75
Connection matrix 80-1
Cycle in a graph 75

Deo, N. 89-90
Derangement 66-7, 94
Degree of a point 77
Descending factorials 13-17, 195
Determinant 53
Difference 6-8
Difference equations (see also
   Recurrence relations) 4-53
   55-7, 73, 195-201, 210
   complete 8, 11-23
   first order 36-7
   homogeneous 8, 10-12, 36-7
   linear 7-35
   partial 48-51
   simultaneous 25-6, 32-4
   variable coefficients 36-42
Difference representation of a
   composition 118
Digraph 74, 96
Dijkstra, E. W. 83, 89-90
Dijkstra's shortest path algor-
   ithm 82-5, 136
Distinct representative (see
   Systems of distinct representa-
   tives)
Distribution 59-60, 94, 203
Downton, F. 52
Dummy variable 24
Dynamic programming 149-54,
   156-7, 161-2

Edges of a graph 74
Eratosthenes sieve method 208

Enumerators (see also Generating functions)
  combinations 56-7, 62
  compositions 69-70
  partitions 70-3
  permutations 55-6
Euler 72, 77
  circuit 75, 77-9, 94
  path 75, 77, 79, 96
Exponential generating functions 63-4, 94

Factorial digits 100-3, 127, 205
Factorial representation 101-3, 107, 109, 123, 204
Feller, W. 89
Female optimal stable solution 183
Ferrers graph 73-4, 94, 202-4
Fibonacci numbers 91, 210
Fike's order for permutations 106-7, 115, 125-8
Filing rules 97-9
Forward differences 6-8, 43-5

Gale, D. 182, 188-9
Gale and Shapley algorithm 185, 187-8, 212
Generating functions (G. F.) (see also Enumerators) 23-5, 34-5, 38-9, 41-2, 52, 60-4, 68-73, 91, 94, 202-3
  exponential 63-4, 94
Golomb, S. W. 156-7, 161
Graphs (linear) 74-88
  complete 174, 191
  directed (digraph) 74, 96
  undirected 74, 79

Hall, Marshall 89, 101, 126 157, 175, 188-9
Hall, Philip 165, 188-9
Hall's algorithm 169-71, 182, 189, 209
Hall's theorem 165-9, 174, 182, 190, 192, 211
Hamiltonian path (cycle) 79
Harary, F. 89-90
Hopcroft, J. E. 156-7
Hungarian method (see also Assignment, classical) 176

Incidence matrix 79-80
Inclusion-exclusion, principle of 64-7, 94, 203
Indicial equation 8 (also throughout chapters 2, 3)
Induction 4

Job-shop scheduling 143-7, 155-7, 160
Johnson, S. M. 125-6

Key 97
Knuth, D. E. 89, 156-7
Kohler, W. H. 156-7
Konig's theorem 173-4, 177-8, 188, 192
Königsberg bridge problem 77, 79, 94

Lawler, E. L. 156-7
Leaves of a tree 133
Length of a path 77
Lehmer, D. H. 125-6, 129, 156-7, 205
Levine, N. 160
Lexicographical ordering 97-102, 105-7, 108-9, 112-13, 115-17, 127-9, 132, 137-8, 157, 159-60, 204, 206
Linear graph (see Graphs)
Linear triplets 32, 196
Lines of a graph 74
Liu, C. L. 54, 89
Loops of a graph 75

MacMahon, P. A. 89
McVitie, D. G. 189
McVitie-Wilson algorithm 185-8, 212
Male optimal stable solution 183, 193-4, 212
Matching 164
Mathematical induction 4
Matrices of zero's and one's (see also Konig's theorem) 164, 173-4, 192
Merging 35, 49-52
Milne-Thomson, L. M. 31, 51
Minimum choice stable solution 184

Multiconnections in a graph 75
Multiplying matrices 161-2

Network 80, 82, 84, 152-4
Number systems 99-103

Optimisation 3, 139-40, 147,
  149-54
Orderings (see also Adjacent mark
  order, Alphabetic order, Fike's
  order, Lexicographical ordering,
  Reverse lexicographical ordering,
  Systematic ordering, Wells'
  order) 97-107

Page, E. S. 52, 90, 125-6
Particular solution 12-23
  for exponentials 18-21
  for polynomials 12-18
Partitions 67-8, 88, 91-4, 132,
  150-2, 159-60, 202-3, 207-8
  conjugate 74
  generation 120-1, 129
  Pascals triangle 57
  self conjugate 74
Paths in a graph 75, 77-9, 82-5,
  152-4
Permutations 5, 31-2, 36, 39-42,
  54-5, 62-4, 66-7, 94, 101-3,
  122-5, 130-2, 137-8, 140-2,
  153-4, 160-1, 196, 204-7
  generation 103-5, 108-16,
    127-9, 157, 204; adjacent
    mark method 109-11, 127;
    derangement method 101-3;
    lexicographical order, (all
    perms) 108-9, (k^{th} perm)
    112-13, 129, 205-6; re-
    peated marks 115-16
  ordering 105-7
Points in a graph 74
Polynomial (see Particular solution)
Polynomial coefficients 37-8

Quaternary sequences 34
Queen's problem 156

Ramsey's theorem 174-5, 188
Random compositions 129,
  206
Random permutations 104-5,
  113-15
Recurrence relations (see also
  Difference equations) 4-7, 94,
  122-5, 129-30, 158, 191, 206
Recursion 121-2, 136
Reduction of order 42-8
Repeated summation 23
Reverse lexicographical order-
  ing 105-6, 119, 121, 127,
  132, 204
Riordan, J. 54, 88-9
Rohl, J. S. 125-6
Rooted trees 86

Salzer, H. E. 160
Scheduling example (see Job-
  shop scheduling)
Searching 131-63
Sedgewick, R. 125-6
Selection sort 114
Seppänen, J. J. 89-90
Shapley, L. S. 182, 189
Shen, M. K. 125-6, 204
Shift operator 6-8, 43-5
Silver, R. 188-9
Sisson, R. L. 156-7
Slings in a graph 75
Spanning tree 76
Spanning tree algorithm 86-8
Sorting 2, 49-51, 154
Stable marriage (see also Gale
  and Shapley algorithm,
  McVitie-Wilson algorithm)
  182-3, 188-9, 193-4, 212-13
Stack 142, 146-7
Steiglitz, K. 156-7
Stirling's approximation 209
Subgraph 76
Sum free sets 156
Summing operations 15-16
Symbolic methods 12-16, 18-19
Systems of distinct representa-
  tives 156, 164-73, 188-91,
  209-11
Systematic ordering 99, 124

'Travelling salesman' problem 3,
79, 104, 150, 153-5, 162
Trees 76, 85-90, 133, 138, 140-2,
144-9, 157, 166, 171, 206-7
Trial solutions 16-18, 20-1, 47-8
199
Trigonometrical functions 20-1
Trotter, H. F. 125-6

Ullman, J. D. 156-7

Vacancies on a line 46-8, 52,
199-200
Vertices in a graph 74

Walker, R. J. 156
Wells, M. 89-90, 125-6, 156-7
Wells' order for permutations 106
Wilson, L. B. 52, 90, 189
Wilson, R. J. 89-90
Wirth, N. 189
Wood, D. E. 156-7